Wireless Power Transfer for Electric Vehicles

Wireless Power Transfer for Electric Vehicles

Editor

Adel El-Shahat

MDPI • Basel • Beijing • Wuhan • Barcelona • Belgrade • Manchester • Tokyo • Cluj • Tianjin

Editor
Adel El-Shahat
School of Engineering
Technology
Purdue University
West Lafayette
United States

Editorial Office
MDPI
St. Alban-Anlage 66
4052 Basel, Switzerland

This is a reprint of articles from the Special Issue published online in the open access journal *Energies* (ISSN 1996-1073) (available at: www.mdpi.com/journal/energies/special_issues/Wireless_Transfer).

For citation purposes, cite each article independently as indicated on the article page online and as indicated below:

LastName, A.A.; LastName, B.B.; LastName, C.C. Article Title. *Journal Name* **Year**, *Volume Number*, Page Range.

ISBN 978-3-0365-6078-6 (Hbk)
ISBN 978-3-0365-6077-9 (PDF)

© 2022 by the authors. Articles in this book are Open Access and distributed under the Creative Commons Attribution (CC BY) license, which allows users to download, copy and build upon published articles, as long as the author and publisher are properly credited, which ensures maximum dissemination and a wider impact of our publications.

The book as a whole is distributed by MDPI under the terms and conditions of the Creative Commons license CC BY-NC-ND.

Contents

About the Editor vii

Alexander Kuehl, Maximilian Kneidl, Johannes Seefried, Michael Masuch, Michael Weigelt and Joerg Franke
Production Concepts for Inductive Power Transfer Systems for Vehicles
Reprinted from: *Energies* 2022, 15, 7911, doi:10.3390/en15217911 1

Niklas Pöch, Inka Nozinski, Justine Broihan and Stefan Helber
Numerical Study on Planning Inductive Charging Infrastructures for Electric Service Vehicles on Airport Aprons
Reprinted from: *Energies* 2022, 15, 6510, doi:10.3390/en15186510 13

Emin Yildiriz and Murat Bayraktar
Design and Implementation of a Wireless Charging System Connected to the AC Grid for an E-Bike
Reprinted from: *Energies* 2022, 15, 4262, doi:10.3390/en15124262 39

Mahmoud Rihan, Mahmoud Nasrallah, Barkat Hasanin and Adel El-Shahat
A Proposed Controllable Crowbar for a Brushless Doubly-Fed Reluctance Generator, a Grid-Integrated Wind Turbine
Reprinted from: *Energies* 2022, 15, 3894, doi:10.3390/en15113894 55

Nishant Patnaik, Richa Pandey, Raavi Satish, Balamurali Surakasi, Almoataz Y. Abdelaziz and Adel El-Shahat
Single-Phase Universal Power Compensator with an Equal VAR Sharing Approach
Reprinted from: *Energies* 2022, 15, 3769, doi:10.3390/en15103769 85

Shamala Gadgil, Karthikeyan Ekambaram, Huw Davies, Andrew Jones and Stewart Birrell
Determining the Social, Economic, Political and Technical Factors Significant to the Success of Dynamic Wireless Charging Systems through a Process of Stakeholder Engagement
Reprinted from: *Energies* 2022, 15, 930, doi:10.3390/en15030930 105

Valerio De Santis, Luca Giaccone and Fabio Freschi
Influence of Posture and Coil Position on the Safety of a WPT System While Recharging a Compact EV
Reprinted from: *Energies* 2021, 14, 7248, doi:10.3390/en14217248 123

Yuan Li, Shumei Zhang and Ze Cheng
Double-Coil Dynamic Shielding Technology for Wireless Power Transmission in Electric Vehicles
Reprinted from: *Energies* 2021, 14, 5271, doi:10.3390/en14175271 133

T. A. Anuja and M. Arun Noyal Doss
Reduction of Cogging Torque in Surface Mounted Permanent Magnet Brushless DC Motor by Adapting Rotor Magnetic Displacement
Reprinted from: *Energies* 2021, 14, 2861, doi:10.3390/en14102861 153

George S. Fernandez, Vijayakumar Krishnasamy, Selvakumar Kuppusamy, Jagabar S. Ali, Ziad M. Ali and Adel El-Shahat et al.
Optimal Dynamic Scheduling of Electric Vehicles in a Parking Lot Using Particle Swarm Optimization and Shuffled Frog Leaping Algorithm
Reprinted from: *Energies* 2020, 13, 6384, doi:10.3390/en13236384 173

About the Editor

Adel El-Shahat

Dr. Adel El-Shahat is an Assistant Professor of Energy Technology, School of Engineering Technology at Purdue University, USA. He is the Founder and Director of Advanced Power Units and Renewable Distributed Energy Lab (A_PURDUE). He holds full-time academic positions at Purdue University, Georgia Southern University, the University of Illinois at Chicago, and Ohio State University in the USA and at Suez University in Egypt. He received his B.Sc. in Electrical Engineering from Zagazig University, Egypt, in 1999; his M.Sc. in Electrical Engineering (Power and Machines) from Zagazig University, Egypt, in 2004; and his Ph.D. degree (Joint Supervision) from Zagazig University, Egypt, and The Ohio State University (OSU), Columbus, OH, USA, in 2011. His research focuses on Modeling, Design, Optimization, Simulation, Analysis, and Control of various aspects such as Renewable Energy Systems; Smart Nano and Micro-Grids; Electric Mobility and Transportation Electrification; Wireless Charging of Electric Vehicles; Climate Change; Electric Vehicles; Special Purposes Electric Machines; Deep Learning Techniques; Distributed Generation Systems; Thermoelectric Generation; Special Power Electronics Converters; Power Systems; Energy Storage and Conservation; and Engineering Education. So far, he has published 10 books, 5 chapters in books, 70 journal papers, 73 conference papers, and 106 other publications with his collaborators and students on topics related to his research interests. Additionally, he has distinguished professional training, and he is a Senior Member of the IEEE and IRED institutions along with 21 professional memberships in other societies. He served as a book editor for 6 books and a reviewer for 9 books. He is a guest editor and editor-in-chief for two international journals. Additionally, he is a reviewer for 35 other international journals. Moreover, he served as an invited conference session chair and reviewer for 33 international conferences along with other community and academic services.

Article

Production Concepts for Inductive Power Transfer Systems for Vehicles

Alexander Kuehl *, Maximilian Kneidl, Johannes Seefried, Michael Masuch, Michael Weigelt and Joerg Franke

Institute FAPS, Friedrich-Alexander-Universität Erlangen-Nürnberg, 91058 Erlangen, Germany
* Correspondence: alexander.kuehl@faps.fau.de

Abstract: The option of wireless energy transmission in electric vehicles can become the main market driver for electric vehicles due to its distinct advantages, such as range, weight, or costs, over conventional conductive charging solutions. In addition to the great potential, which different research work and realized systems have already shown, there are new requirements for the associated production networks in the automotive industry which must be addressed at an early stage. Furthermore, no solutions currently exist for the industrial production of these components. This paper presents the main components for the feasibility of wireless power transmission in electric vehicles. In addition, the required value chains and processes for the new components of the inductive power transfer systems, and the final assembly for induction coils, which has been developed at the FAU, will be presented. These include the developing of a winding process on a 15-axis special machine, ultrasonic crimping of the litz wire ends, and vacuum potting.

Keywords: charging automation; electric vehicles (EVs); wireless charging; wireless power transfer (WPT); production; automation; inductive power transfer (IPT); manufacturing

Citation: Kuehl, A.; Kneidl, M.; Seefried, J.; Masuch, M.; Weigelt, M.; Franke, J. Production Concepts for Inductive Power Transfer Systems for Vehicles. *Energies* **2022**, *15*, 7911. https://doi.org/10.3390/en15217911

Academic Editor: Adel El-Shahat

Received: 31 August 2022
Accepted: 14 October 2022
Published: 25 October 2022

Publisher's Note: MDPI stays neutral with regard to jurisdictional claims in published maps and institutional affiliations.

Copyright: © 2022 by the authors. Licensee MDPI, Basel, Switzerland. This article is an open access article distributed under the terms and conditions of the Creative Commons Attribution (CC BY) license (https://creativecommons.org/licenses/by/4.0/).

1. Introduction

Over the past years, electro mobility has become an important topic in emission free transportation plans. In combination with renewable energies, it is even possible to keep vehicles carbon neutral. These advantages have triggered the interest in electric vehicles in recent years, but the technology has always been expensive. Significant factors leading to low user acceptance include the large and expensive batteries in short-range vehicles and slow charging solutions. Wireless power transfer (WPT) is focused on solving these current problems, such as the inconvenient plug-in charging process. Modern systems have already demonstrated the potentials of wireless energy transfer in cars [1,2]. However, there are currently no solutions for the industrial production of these components.

Within this paper the results of a feasibility study of contactless charging will be introduced, followed by the production processes for inductive power transfer (IPT) systems as well as automation concepts, which have been developed at the Institute for Factory Automation and Production Systems (FAPS) of the Friedrich-Alexander-Universität Erlangen-Nürnberg (FAU).

2. Components and Types of Wireless Power Transfer

The idea of wireless power transfer has existed since the late 19th century [3]. Currently, a variety of different systems exists with transmitting powers from 10^0 W (e.g., mobile phones) up to 10^5 W to supply power into electrical trains.

Particularly in the field of medical and consumer products, WPT systems are proven and widely used. About fifty years ago, the first WPT artificial implanted heart have been published [4]. At the end of the 1970s, WPT charged electric toothbrushes [5] kicked off the success story of wireless power transfer in the consumer industry. The integration in the industrial production was triggered in the 1990s by automated guided vehicles (AGV), of which wireless charging was an improvement because of its greater robustness and lower maintenance [6].

WPT systems for electric vehicles (EV) have been a logical extension of this. They have already existed in research projects since the 1970s [7] and have spurred new interest as recent improvements in material and power electronics have arisen. In a direct comparison, automotive applications have requirements that are more stringent in the points of air gaps, power level and power density as well as safety (several nations have defined safety limits, values and zones for the electromagnetic field, e.g.,), compared to AGV applications [6]. In 2021, Momentum Dynamics issued a declaration of CE conformity for its up to 300 kW wireless charging solutions [8].

2.1. WPT Technologies

The use cases of wireless charging are listed in Table 1:

Table 1. Different WPT technologies for EVs [9].

	Stationary WPT	Semi-Dynamic WPT	Dynamic WPT	
Power rates	3.7–22 kW	3.7–22 kW	>20 kW	
Examples	Home, job or shopping center	Traffic lights, taxi ranks, bus stations	Along gas stations, on motorways	Charging lanes on motorways

2.1.1. Stationary Charging

Private vehicles are usually stationary objects. The typical daily drive time of a private car in Germany is 32 min, and most trips are shorter than 40 km (~25 mi) [10]. Average European distances are half as long as in the USA [11]. Therefore, charging with about 3.7 kW for vehicles with approximately 20 kWh energy storage capacity are sufficient. EV drivers, who do not have the possibility to charge their own equipment, depend on public charging infrastructure solutions. To promote socio-economic acceptance of electromobility, the range has to grow, while charging times have to be reduced by the use of fast charging batteries and charging stations. Otherwise, public infrastructure must be profitable and integrable [12].

2.1.2. Semi Dynamic Charging

A further step is charging in idle positions or slow-driving sections, e.g., taxi stands, traffic light or traffic-calmed sectors, where conductive charging is ineffectual because of the short cycles. These sections can be equipped with switched IPT pads, which are enabled when the vehicles of customers cross these pads. The technology enables energy transfer in cars during waiting periods of conventional urban driving profiles. The benefits are higher ranges and no need of special charging points. Besides economic problems such as higher installation and maintenance costs compared to home based technology, new traffic scenarios have to be researched, evaluated, defined and coordinated with transport authorities [13,14].

2.1.3. Dynamic Charging

The next, more holistic step is supplying vehicles with energy while driving at cruising speed, which eliminates problems like limited range and therefore allows a reduction of battery capacity.

The first major academic project was coordinated by the "Partners for Advanced Transit and Highways" (PATH), led by UC Berkeley [7]. The consortium equipped an electric bus with a pickup system and used a street with a track system (see Figure 1).

Figure 1. Roadway track system of the KAIST project.

The results of the project have proven that inductive charging while driving is technically working but limited to materials and electronics that have been available at this time. However, the costs of the installation and maintenance of infrastructure must be reduced. Within the last decades many other test tracks have been installed, such as by the KAIST in South Korea for buses, by Bombardier in Germany for tramways, or by Electreon in Sweden and Germany for trucks [15–17].

Nevertheless, there have been no efforts to industrialize the production of the components of this technology and their integration into traffic routes.

2.2. Components of WPT

A WPT system in the variant of IPT consists of a primary and a secondary coil, as well as the associated power electronics and sensors. The primary system delivers the energy and might be installed in the infrastructure (see Figure 1). The secondary system receives the energy to supply the vehicle and is shown in Figure 2. The main components are the coil and the ferrite blocks, which are insulated by a resin as well as the connection cable.

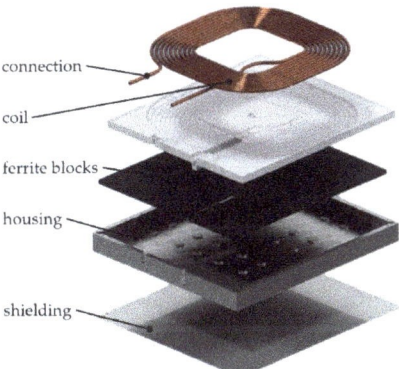

Figure 2. Composition of an IPT module.

Depending on the type of use, charging pads have different coil designs, which result in different production challenges. To give some examples, there are single-sided and double-sided pad structures with e.g., circular, solenoid, H-formed, D-formed, and DDQ-formed coil designs [18,19].

3. Feasibility of Contactless Charging of EVs

According to the state of the art, several hundred kW have already been transferred to moving vehicle systems (see Figure 3) under certain conditions. At a distance of 17 cm, the

South Korean consortium OLEV (On-line Electric Vehicle) has already been shown to have an efficiency of more than 71% at 17 kW, according to official figures. The first test systems established were able to comply with the ICNIRP limit recommendations of 6.25 µT in the relevant areas around the vehicle [15].

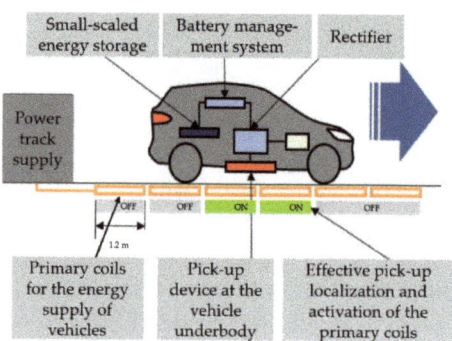

Figure 3. System for WPT in moving cars [20].

Further research is needed, among other things, to increase efficiency and reduce vehicle and infrastructure costs, on the one hand through innovative energy transmission topologies and, on the other hand, through the efficient production and integration of these systems in vehicles and on the road. In particular, automated manufacturing processes are needed to reduce system costs, as very high relative costs are created in the initial migration stages to refinance the infrastructure.

4. Production Concepts for Inductive Power Transfer Pads

The current state of IPT pad production is characterized by small lot sizes and manual work. This often results in fluctuations of quality, which can induce lower power transfer efficiency, damage, or reduced service life expectancy. On the other hand, a higher demand leads to higher lot sizes and automated processes. In the following chapter, process chains will be explained and suitable processes shown.

4.1. Process Chain for IPT Pads

Charging pads consist of different components and materials which are manufactured, handled, and assembled in several production steps (Figure 4 (right)). The first step is the preproduction of the base parts and the housing. The functions of these parts are to position coils, electronics, and ferrites and to encapsulate the system (mechanically, chemically, thermal, etc.).

The following steps (winding and contacting of the high-frequency (HF) litz wires) will be discussed in Sections 4.2 and 4.3.

In another step, the mentioned ferrites and electronic packages must be assembled. The challenges of this production step include the careful handling and most precise dropping of the ferrites, which have tight tolerances, because the position influences the magnetic field.

With rapidly increasing demand for solutions for wireless charging systems, adaptive value chains with a gradually increasing degree of automation have to be created. To reach short cycle times, it is important to process a wide variety of materials safely and with high process stability. The multitude of fields of action for wireless charging systems from a production and material perspective also requires interdisciplinary manufacturing process development. With an initially low level of integration of the electronics in the charging modules, partial pre-assembly and separate delivery of the individual components (coil module, compensation, rectification, etc.) may be necessary, which is accompanied by complex final assembly. Since, above all, the logistics processes and the associated costs have a significant influence on the further design of the modules, a higher level of integration is to be aimed at in later stages of development. It is also conceivable to integrate the systems into the battery modules, which for reasons of logistics (costs and safety of transport) tend to be produced very close to the final assembly plants.

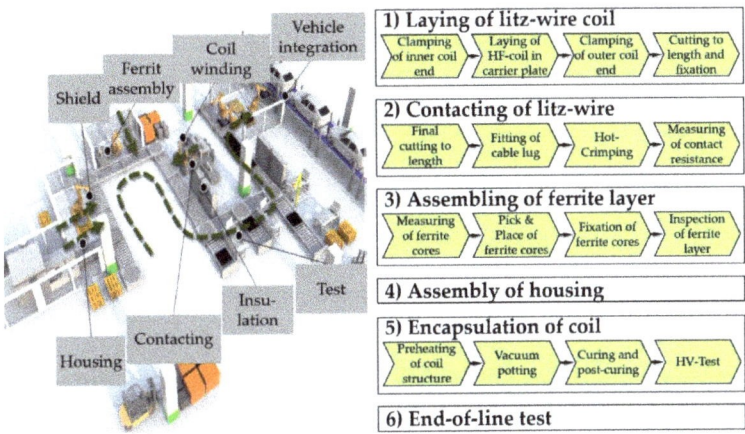

Figure 4. CAD model of a value-added chain [21] (**left**) and process chain of the IPT pad production (**right**).

Additional advantages arise in the case of integration of the electronics close to the target site by minimizing the high-voltage cabling expenditure and the synergetic use of already integrated cooling concepts to be able to ensure the required heat dissipation, especially in higher performance systems.

In the case of highly automated processing and handling of magnetic field guidance materials, e.g., ferrites, the risk of breakage of the brittle material must be considered. In addition to the development and construction of suitable litz-wire laying techniques for optimized winding patterns for better heat dissipation, methods for process-stable, large-series-capable contacting of the several hundred individual wires are to be developed for the high-frequency strands used. Current methods of thermal stripping and contacting via tin baths are not suitable for large series and automotive applications. For the following processes, appropriate buffer systems must be used to meet the required conditions for the curing process of insulation and potting materials and to ensure an approximately constant flow of the material.

In the vehicle final assembly plant, the charging modules are installed in the underbody of the vehicle. At present, there are no vehicle concepts that have been completely designed for a contactless charging option, especially with regard to the body structures including the battery module in the underbody area. Optimal integration, ideally in the center of the underbody to ensure interoperability between vehicle models while preserving mechanical, electrical, chemical and thermal boundary conditions, will only be possible in future generations of vehicles.

Within the next sections, the focus of the IPT pad production will be set on the winding, contacting and insulating technologies of HF litz wires.

4.2. Winding Technologies

The difficulty in automated production of coils is represented by the various coil parameters.

Particularly worthy of mention here are the 3D contour, the structure as a multi-coil system, multipole coils, systems with several HF-litz wires, the winding distance, various radii, winding tolerances, wire intersections and the positioning of the coil ends.

4.2.1. Processing of HF-Litz Wires

The starting point is the use of already stranded wires. The supply path of the strands begins with the supply from storage reels or ones that are cut-to-length. The strands are then pulled through the feeder via a position measuring system and the dancing roll and are then guided to the placement point via guide rollers and the gripper.

4.2.2. Stabilization of Coils

Complex coil constructions have to be inherently stable for several reasons: stability facilitates logistics (e.g., transport of preformed coils from the winding machine to assembly in the housing) and contributes to protecting the sensitive materials in the pads from damage. It simplifies the positioning of components and helps to maintain even complex shapes such as correctly distributed winding distances or to reduce process time. Depending on how the coils are possibly fixed in the pad, e.g., with full encapsulation, they may only have to be self-stable for a short period of time.

There are various options for coil carrier materials and fastening methods that can be divided or combined into mechanical (sewn on textile, staples, base plates with hooks or grooves, rigid coil formers) and adhesive (liquid or sprayed adhesives, adhesive foils, or self-adhesive cables). Various concepts were tested for this purpose. Base plates with grooves enable the high-precision placement of the winding ends. The combination of this plate and a laying tool with, for example, a tool for feeding adhesive strips also enables accurate placement and fixation of the winding. Another tested system includes a vacuum gripper that grips self-adhesive strands and places them on a flat plate. By adding a heat source (hot air or laser), the cables can be fixed directly to the areas that are important for the shape.

4.2.3. Winding of HF-Litz Wires

There are three basic concepts to perform relative movements between the laying tool and the base plate, which are also realizable with robots [22]. While manually produced demonstrators usually have certain imperfections and higher manufacturing tolerances, automated test setups are usually expensive and inflexible. The use of a universal and flexible robot kinematics in combination with the realization of a cad-cam chain makes it possible to use a robot for demonstrator assembly as well as for serial production.

For validation, a layout was created which includes different radii as well as parallel windings and winding intersections. In addition, various installation tests were carried out in which the strands (material, diameter, structure), the shunt radii and the setting radii can be varied. These tests make it possible to draw conclusions about the reproducibility of a pad during the electromagnetic design phase.

For example, an R14 radius can no longer be produced with a 768 × 0.1 mm HF strand, or a radius of 1 mm with an angle of attack of 90 degrees, but only with insulated strands (e.g., a 300 × 0.2 mm baked varnish strand).

Within the framework of research, basic production concepts have been realized. In order to achieve high-quality IPT systems, not only the basic production concepts but also the material behavior must be considered in detail. To reflect critical material behaviors, selected production concepts were therefore implemented and scrutinized. Overall, it became evident that self-bonding wires can be processed more easily without external insulation. However, HF litz wires without external insulation have the disadvantage of a lower cohesion of the individual strands and less protection against mechanical damage. As shown, problems can occur in the winding process with tighter curves if the wires are twisted and deformed in the winding direction.

Complex coil constructions require external insulation to keep the wires in shape and ensure sufficient electrical insulation, even in difficult environments. However, inelastic adhesive tapes make stranded wires very stiff and therefore difficult to wind, especially in tighter curves, and the tape needs to be removed before contact to avoid residues and vapors. Outer insulation with silk offers the best compromise, namely high HF stranded wire stability and good protection against mechanical damage, while at the same time offering flexibility. A disadvantage is the need to wipe the silk before the contacting process to avoid residues and vapors.

In summary, the tests described above can be used to determine production-relevant parameters in order to assess the suitability of HF stranded wires for winding as well as the necessary winding and wire laying tools and to produce prototypes.

4.3. Contacting Technologies

Due to the high number of insulated single wires of a high-frequency litz wire, the process step of contacting is a challenging process. In order to enable an electrical and mechanical connection of the high-frequency litz wire with contact elements such as cable lugs or terminals, the primary insulation of the single wires must be removed. The high number of single wires precludes mechanical stripping prior to the actual contacting process. Chemical stripping processes are also extremely challenging and do not allow a sufficient stripping. However, in order to enable the contacting of primary-insulated litz wires, combined contacting processes can be used. These realize the necessary stripping process directly during the actual contacting. The hot and ultrasonic crimping processes are particularly suitable for this purpose. In both processes, the contacting is realized by a crimping process with tubular cable lugs. On the other hand, the primary insulation is removed thermally. While the hot crimping process utilizes the resistance heating of electrodes and tubular cable lugs for this process, ultrasonic crimping uses only the damping of ultrasonic oscillations. The resulting conversion of oscillation energy into thermal energy also enables a thermal stripping. Furthermore, high-frequency litz wires can be stripped by means of an ultrasonic compacting process. Again, the damping of mechanical oscillations leads to thermal stripping. At the same time, the tooling system used produces a defined node cross section with welded single wires. The compacted high-frequency litz wire can subsequently be welded onto contact terminals using the ultrasonic welding process. A direct welding of high-frequency litz wires with terminals is rather difficult to implement since the primary insulation hinders the welding process [23].

4.4. Insulation Technologies

After the connection, the pad must become robust, electrically insulated and protected from external influences (water, dust, salt, etc.).

The insulating matrix around the high-frequency litz wire coil is a combination of the primary and secondary insulation. While the term secondary insulation clearly refers to resin encapsulation, the primary insulation can be divided into the litz wire enamel (standardized according to IEC 60317, IEC 60851) and the wrapping made of natural or synthetic materials.

In the context of this study, the primary and secondary insulation differ significantly in terms of production and function. For the primary insulation, the individual wires are covered with a thin, electrically insulating enamel. The additional wrapping of the high frequency litz wires is done by banding, extrusion or spinning with natural silk, nylon, polyamide, polyester, polyethylene or polyimide. Despite the large number of individual conductors, this leads to a high degree of cohesion of the strand and to a high mechanical stability during the laying of the coil. The secondary insulation ensures the protection of the electrical components against mechanical and chemical environmental impact and the dissipation of generated heat. A dielectric potting material should be selected to keep eddy current losses low and to prevent electrical breakdown [24–26].

5. Implementation on a Concrete Demonstrator

Within the E | Profil project funded by the BMWi and DLR, a demonstrator was set up at FAU. The design is specified in accordance with WPT3 SAE J2954 [27]:

- WPT Power class, 11 kW
- Coil-to-Coil efficiency > 96%

- Dimension: ~300 mm × 300 mm (based on SAE2954 (WPT3 Z1)) [27]
- Input: 11.1 kVA and 400 V
- Stranded HF litz wire

The selection and implementation of the processes for the project demonstrator will be described in the following sections.

5.1. Laying of the Litz Wire

After the controlled uncoiling of the stranded wire from the supply reel, the stranded wire is laid with a coiling tool. For this purpose, a specific guide and wire tension control for stranded wire have been integrated into a custom universal 15-axis winding machine. The wire is guided to the deposit point and deposited in a rotating spool carrier with guide grooves which remains in the subsequent spool module (see also Figure 5 (left)).

Figure 5. Manufacturing cell for flat coil winding (**left**), ultrasonic welding system and contacted litz wires (**right**).

5.2. Contacting

Within the scope of the production of the demonstrator, various contacting solutions for contacting the high frequency litz wire are being evaluated. These include the already mentioned technologies of hot and ultrasonic crimping, but also processes that enable a contacting on contact terminals. The primary insulation used complicates terminal contacting in particular. Nevertheless, it has been shown that torsional ultrasonic welding can also be used to weld primary-insulated litz wires directly onto contact terminals. In a two-stage process, the litz wire is compacted, stripped and welded onto the terminal. However, necessary gap dimensions between the tools as well as burning residues require further optimization of the welding process. Furthermore, the ultrasonic crimping process shows to be an energy-efficient alternative contacting technology for tubular cable lugs. However, there is also a demand for further optimization of this process in order to ensure that the technology is suitable for series production. The hot crimping process has proven to be an extremely robust and well-suited process. The achievable contacting quality is subject to only minor fluctuations. By slightly adjusting the process parameters, it is also possible to join taped and PAI primarily insulated high-frequency litz wires to tubular cable lugs. Therefore, the hot crimping process is used as a reference contacting process in this research project. However, the extremely promising torsional ultrasonic crimping and welding process will be further optimized for future applications. Figure 5 (right) shows, in addition to exemplary test samples, the ultrasonic welding system used and various contacts produced.

5.3. Insulating

A dielectric potting material should be selected as secondary insulation to keep eddy current losses low and to prevent electrical breakdown. Epoxy resin is characterized by excellent product and process properties and is therefore widely used in encapsulation processes [24].

Several sub-process steps are necessary to prepare a self-leveling, thermosetting, dielectric molding mass. Figure 4 gives a detailed view of the insulation process and its integration into the IPT process chain. Each sub-process step is characterized by different process parameters.

The process for encapsulating the electrical components consists of the resin preparation, the dosage and the mixing of the resin and hardener, the potting, and finally the curing of the molding mass. During the preparation phase, the resin and the hardener get tempered and dried to reduce moisture and set the process relevant viscosity for the mixing and potting. The mixing step is needed to homogenize the resin and hardener to a reactive molding mass. Within this step, the polymerization begins and the potlife must be monitored carefully. The potting phase is the actual insulation process of the IPT, where the molding mass is dosed via the casting nozzle into the coil and carrier. By processing in a vacuum chamber (see Figure 6a), a higher degree of litz wire impregnation can be achieved [26]. After the potting, the coil is transferred to the curing stage, where the crosslinking and therefore the viscosity of the resin material increases. By processing in a convection oven, the material specification such as the glass transition temperature and the degree of crosslinking can be set to defined values. By varying these, the challenges in the processing of resin-based insulation materials for wireless power transfer applications can be evaluated in more detail.

The studies of M. Kneidl et al. [26], summarized the most important process parameters of the complete encapsulation process, which result in a good product quality of the insulation system. By analyzing the process chain, the overall process time is mostly determined by the curing step of the insulation resin. Therefore, the curing time can be reduced by 75% with increased temperatures of up to 80 °C. Another positive effect due to higher material and curing temperatures is a better impregnation of the high-frequency litz wires, due to the lower viscosity of the resin. This results in optimized dielectric properties. Further on, optimizing the insulation quality by using a vacuum process is dependent on the type of primary insulation, which is used for the litz wires [26].

5.4. Assembly of the Ferrites

For the assembly of the ferrites, the following process steps result in the system setup, shown in Figure 6b). The system components are mounted on the workstation base with numerous mounting options. The components include the cycle chains as provisioning tools for material replenishment, a cobot as the central handling device with the vacuum gripping system, and the measuring station with an optical 2D industrial camera.

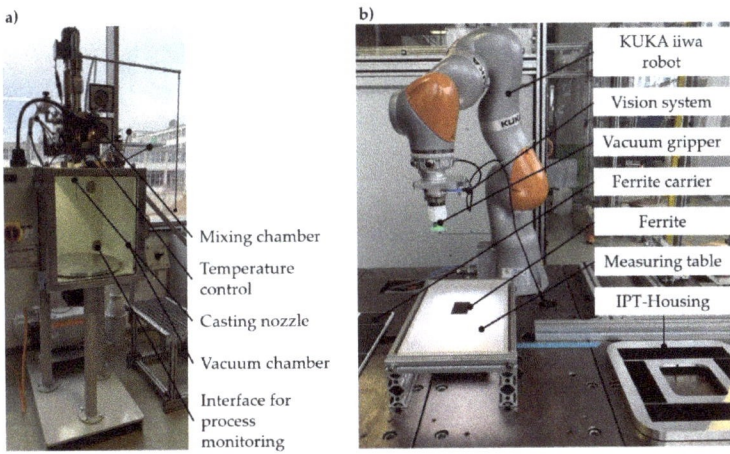

Figure 6. Vacuum potting machine (**a**), robot-based ferrite structure assembly (**b**).

In the first process step, the fixture of ferrites in the charge pad housing is prepared. The fixation can be realized by adhesives, tapes or varnishes.

The next process step is the individual conveying of a ferrite through the system. In interaction with the vacuum end effector, the robot arm takes a ferrite tile from the workpiece carrier and conveys it to the measuring table. For the robot to pick up the ferrite, the control program must know the information about the arrangement and current stock of materials in the workpiece carrier. A constant supply of compressed air for the vacuum gripper is necessary so that a secure fixation can be maintained permanently. The most important tasks are the safe holding of the workpiece and the fast and collision-free movement through the process plant. All stations must logically be accessible in the COBOT's workspace.

For high process speeds, the travel distances must be kept as short as possible and non-linear. This means that speed-optimized conveying additionally contributes to economical production. However, adapted and linear movements should be used for the "pick-up", "measure" and "assemble" process stations. This increases the process reliability and precision of the work steps and ensures collision-free pickup and assembly processes.

The optical 2D industrial camera records the geometric dimensions, any damage present, and the orientation of the current ferrite on the backlit measuring table. The COBOT moves to a defined position in the detection range of the camera. There it remains in a waiting position until the inspection is completed. The dimensions in length and width of the ferrite, as well as its position on the vacuum gripper, are recorded with a certain accuracy. These data can be evaluated and interpreted with suitable software. The *actual* values are compared with the *target* values and manufacturing tolerances from the data sheets. In addition to the deviations of the ferrite dimensions, the deviations from the ideal position of the tile on the gripper is also checked.

When picking up from the blister, twisting or displacement of the breakpoint could occur due to incorrect orientation of the ferrite. All captured and determined data is fed back into the process and included for the assembly process. After being conveyed from the measuring station, the ferrite reaches the mounting station. The COBOT moves to the planned position for the current tile above the housing. The correct position is defined to the robot controller using the coordinate system for the layout according to the loading pad design. Using the information from the survey station, it is known whether the current ferrite has deviations in its manufacturing dimensions and in its position on the gripping system. If this is the case, the robot controller can make the necessary corrections. This means that the ferrite is brought into a suitable position by the movements of the COBOT. The correct positioning of the ferrite tile is thus achieved. However, the accuracy of this result depends on the quality of the measurement data and the robot accuracy. The ferrite is over its assigned place and is ready for the curing of the chosen fixation.

6. Test Technologies

A first test bench is used for recording the efficiency, the offset, the temperatures and the EMC emissions of the pad. Initially, various concepts were developed for recording the above-mentioned variables, whereby the determination of the efficiency or the associated measurement of current and voltage proved to be the most difficult task. Due to the high frequencies of the voltage at the charging pads, it was not possible to use the most common measuring devices. A further challenge was the necessary galvanic isolation, as few measuring systems available on the market have one.

For this reason, completely galvanically isolated measurement electronics was developed and implemented on a 3-Axis linar production cell. In order to detect possible temperature changes in the charge pads or other components of the contactless power transmission system, a temperature measuring system was designed and developed. In order to be able to carry out EMC measurements, an EMC probe was integrated into the test bench. Through the produced serial cable and the integration into the created LabView, the probe can be used for measurement. At the serial output, it supplies only field strengths averaged over all spatial axes, but this is sufficient for an evaluation of the measured field strengths. IPT systems, as the limit values to be complied with by legislators, are determined in the same way.

In a second step, an EOL test was build up. It is able to measure the quality factor of the coil or the transmission efficiency in power transmission mode.

7. Conclusions

In this paper, the principle of wireless vehicle charging and the associated process chain were presented. Using the example of a demonstrator set up in the E|Profil project, the process chain for manufacturing a secondary system was presented. The next step is the integration of the charging technology into the infrastructure. As part of the E|Road project launched at FAU in summer 2021, the necessary production steps for concrete roads will be investigated. Important research aspects here are:

- Design of the coil modules suitable for production
- Automation-compatible processes for manufacturing the coil modules
- Integration of the coil modules into the roadway using precast concrete construction methods
- Construction of an electrified test track on the duraBASt demonstration site

The completion of this extensive research work is intended to demonstrate that inductive charging technology is a feasible solution for the mobility of the future.

Author Contributions: Conceptualization, writing—original draft preparation and supervision A.K.; processes, writing—review and editing M.K., J.S., M.M. and M.W.; review and supervision, J.F. All authors have read and agreed to the published version of the manuscript.

Funding: This work was supported by the German Ministry for Economic Affairs (BMWi) and DLR by founding the project E|Profil under Grant FKZ 01MV18003A.

Informed Consent Statement: Not applicable.

Data Availability Statement: Data of the project and to the machines is published online at www.faps.fau.de, accessed on 30 August 2022.

Conflicts of Interest: The authors declare no conflict of interest.

Abbreviations

AGV	Automated Guided Vehicles
EMC	Electromagnetic Compatibility
EOL	End Of Line
EV	Electric Vehicle
ICNIRP	International Commission on Non-Ionizing Radiation Protection
IPT	Inductive Power Transfer
HF	High Frequency
OLEV	On-Line Electric Vehicle
PAI	Aolyamideimide
PATH	Aartner for Advanced Transit and Highways
WPT	Wireless Power Transfer

References

1. Perrin, J. Inductive Charging of Electric Vehicles: A European Perspective. In Proceedings of the Conference on Electric Roads & Vehicles, Park City, UT, USA, 5 February 2013.
2. Jeong, S.; Jang, Y.J.; Kum, D.; Lee, M.S. Charging Automation for Electric Vehicles: Is a Smaller Battery Good for the Wireless Charging Electric Vehicles? *IEEE Trans. Automat. Sci. Eng.* **2019**, *16*, 486–497. [CrossRef]
3. Hutin, M.; Leblanc, M. Transformer System for Electric Railways. US Patent 527857, 23 October 1894.
4. Schuder, J.C. Powering an Artificial Heart: Birth of the Inductively Coupled-Radio Frequency System in 1960. *Artif. Organs* **2002**, *26*, 909–915. [CrossRef] [PubMed]
5. Roszyk, L.; Barnas, L. Hand held battery operated device and charging means therefore Publication. US Patent 3840795, 8 October 1974.

6. Covic, G.A.; Boys, J.T. Modern Trends in Inductive Power Transfer for Transportation Applications. *IEEE J. Emerg. Sel. Topics Power Electron.* 2013, *1*, 28–41. [CrossRef]
7. Systems Control Technology Inc. *Roadway Powered Electric Vehicle Project Track Construction and Testing Program Phase 3D*; California PATH Research Paper UCB-ITS-PRR-94-07; Institute of Transportation Studies, University of California: Berkeley, CA, USA, 1994.
8. Momentum Wireless Power. *Momentum Dynamics' Wireless Charging Solution Achieves CE Mark*; Momentum Wireless Power: Malvern, PA, USA, 2021.
9. Risch, F. *Planning and Production Concepts for Contactless Power Transfer Systems for Electric Vehicles*; Meisenbach: Bamberg, Germany, 2014; ISBN 978-3-87525-369-6.
10. Follmer, R.; Gruschwitz, D.; Jesske, B.; Quandt, S. *Mobilität in Deutschland 2008 (MiD): Ergebnisbericht: Struktur—Aufkommen*; FE-Nr. 70.801/2006; Bundesministeriums für Verkehr, Bau und Stadtentwicklung: Bonn, Germany; Berlin, Germany, 2010.
11. Christensen, L.; Fosgerau, M. Impacts from Different Land-Use Strategies on Travel Distances. In Proceedings of the European Transport Conference, Homerton College, Cambridge, UK, 9–11 September 2002.
12. Wechlin, M. Vorrichtung zur Induktiven Übertragung Elektrischer Energie. PCT/EP2013/052019, 1 February 2013.
13. Sithinamsuwan, J.; Hanajiri, K.; Hata, K.; Imura, T.; Fujimoto, H.; Hori, Y. Stop Position Estimation for Automatic Stop Control of Electric Vehicle in Semi-Dynamic Wireless Charging System. In Proceedings of the 2019 IEEE International Conference on Mechatronics (ICM), Ilmenau, Germany, 18–20 March 2019; IEEE: Piscataway, NJ, USA, 2019; pp. 590–595, ISBN 978-1-5386-6959-4.
14. Jaguar Land Rover. *Jaguar I-Pace Electric Taxis on World's First Wireless High-Powerd Charging Rank*; Jaguar Land Rover: Mahwah, NJ, USA, 2020.
15. Suh, I.-S.; Cho, D.-H.; Franke, J.; Hong, S.-M.; Jung, S.-K.; Lee, B.-S.; John, M.M.; Risch, F.; Naoki, S.; Turki, F. *Wireless Charging Technology and the Future of Electric Transportation*; SAE International: Warrendale, PA, USA, 2015; ISBN 9780768081534.
16. Lee, B.S.; Hong, S.M. Application of SMFIR to Trains. In *The On-Line Electric Vehicle: Wireless Electric Ground Transportation Systems*; Suh, N.P., Cho, D.H., Eds.; Springer: Cham, Switzerland, 2017; ISBN 978-3-319-51183-2.
17. Electreon Wireless. Electreon Smart Road Gotland: The World's First Wireless Electric Inter-City Road System, Charging an Electric Bus and an Electric Heavy Duty Truck. Available online: https://www.electreon.com/projects-gotland (accessed on 24 January 2022).
18. Risch, F.; Günther, S.; Bickel, B.; Franke, J. Flexible Automation for the Production of Contactless Power Transfer Systems for Electric Vehicles. In Proceedings of the 3rd International Electric Drives Production Conference (E|DPC), Nuremberg, Germany, 29–30 October 2013; IEEE, Ed.; IEEE: Piscataway, NJ, USA, 2013; pp. 485–491, ISBN 9781479911028.
19. Budhia, M.; Boys, J.T.; Covic, G.A.; Huang, C.-Y. Development of a Single-Sided Flux Magnetic Coupler for Electric Vehicle IPT Charging Systems. *IEEE Trans. Ind. Electron.* 2013, *60*, 318–328. [CrossRef]
20. Risch, F.; Kühl, A.; Franke, J. *Machbarkeitsstudie zum Kontaktlosen Laden von Elektromobilen (E|ROAD): Feasibility Study for Contactless Charging of Electric Vehicles (E|ROAD)*; Abschlussbericht AZ-918-10; Bayerische Forschungsstiftung: Erlangen, Germany, 2010.
21. Risch, F.; Günther, S.; Franke, J. Wertschöpfungsketten für kontaktlose Ladesysteme: Konsequenzen der kontaktlosen Energieübertragung in Elektrofahrzeuge für automobile Wertschöpfungsketten. *Ind. Manag.* 2012, *28*, 45–48.
22. Weigelt, M.; Masuch, M.; Mayr, A.; Kühl, A.; Franke, J. Automated and Flexible Production of Inductive Charging Systems as an Enabler for the Breakthrough of Electric Mobility. In *Advances in Production Research, Proceedings of the 8th Congress of the German Academic Association for Production Technology (WGP), Aachen, Germany, 19–20 November 2018*; Springer: Cham, Switzerland, 2019; pp. 300–309.
23. Seefried, J.; Riedel, A.; Kuehl, A.; Franke, J. Challenges and Solutions for Contacting Insulated Litz Wire Structures in the Context of Electromechanical Engineering. In *Production at the Leading Edge of Technology*; Behrens, B.-A., Brosius, A., Drossel, W.-G., Hintze, W., Ihlenfeldt, S., Nyhuis, P., Eds.; Springer International Publishing: Cham, Switzerland, 2021; pp. 466–475. ISBN 978-3-030-78423-2.
24. Nategh, S.; Barber, D.; Lindberg, D.; Boglietti, A.; Aglen, O. Review and Trends in Traction Motor Design: Primary and Secondary Insulation Systems. In Proceedings of the 2018 XIII International Conference on Electrical Machines (ICEM), Alexandroupoli, Greece, 3–6 September 2018; pp. 2607–2612.
25. Sullivan, C.R. Optimal Choice for Number of Strands in a Litz-Wire Transformer Winding. *IEEE Trans. Power Electron.* 1999, *14*, 283. [CrossRef]
26. Kneidl, M.; Masuch, M.; Rieger, D.; Kuhl, A.; Franke, J. Processing influences of resin-based insulation materials for wireless power transfer applications. In Proceedings of the 2020 IEEE Conference on Electrical Insulation and Dielectric Phenomena (CEIDP), East Rutherford, NJ, USA, 18–30 October 2020; IEEE: Piscataway, NJ, USA, 2020; pp. 551–555, ISBN 978-1-7281-9572-8.
27. Hybrid—EV Committee. *J2954_201904: Wireless Power Transfer for Light-Duty Plug-in/Electric Vehicles and Alignment Methodology*; SAE International: Warrendale, PA, USA, 2019.

Article

Numerical Study on Planning Inductive Charging Infrastructures for Electric Service Vehicles on Airport Aprons

Niklas Pöch [1,2,†], Inka Nozinski [1,2,†], Justine Broihan [1,†] and Stefan Helber [1,*,†]

1. Department of Production Management, Leibniz University Hannover, 30167 Hannover, Germany
2. Cluster of Excellence SE²A—Sustainable and Energy-Efficient Aviation, Technische Universität Braunschweig, 38106 Braunschweig, Germany
* Correspondence: stefan.helber@prod.uni-hannover.de; Tel.: +49-51-1762-5650
† These authors contributed equally to this work.

Abstract: Dynamic inductive charging is a contact-free technology to provide electric vehicles with energy while they are in motion, thus eliminating the need to conductively charge the batteries of those vehicles and, hence, the required vehicle downtimes. Airport aprons of commercial airports are potential systems to employ this charging technology to reduce aviation-induced CO_2 emissions. To date, many vehicles operating on airport aprons are equipped with internal combustion engines burning diesel fuel, hence contributing to CO_2 emissions and the global warming problem. However, airport aprons exhibit specific features that might make dynamic inductive charging technologies particularly interesting. It turns out that using this technology leads to some strategic infrastructure design questions for airport aprons about the spatial allocation of the required system components. In this paper, we experimentally analyze these design questions to explore under which conditions we can expect the resulting mathematical optimization problems to be relatively hard or easy to be solved, respectively, as well as the achievable solution quality. To this end, we report numerical results on a large-scale numerical study reflecting different types of spatial structures of terminals and airport aprons as they can be found at real-world airports.

Keywords: dynamic wireless charging; electric vehicles; airport apron; airport infrastructure planning; electric busses

1. Introduction

A growing number of airports are electrifying their apron vehicle fleets to meet goals for climate-neutral airports (Bopst et al. [1], Interreg CENTRAL EUROPE [2], Flughafen München GmbH [3] and Royal Schiphol Group [4]). At Stuttgart Airport, for example, 40% of apron vehicles are equipped with electric drives and apron buses are already exclusively electrically powered (Bulach et al. [5]). Conductive charging is the state-of-the-art technology for charging these vehicles. However, this technology results in long downtimes due to vehicles charging and requires large batteries. A potential option for charging the vehicle batteries is dynamic inductive charging: Vehicles are wirelessly charged while in motion on a charging track installed below the road surface. This technology can substantially reduce downtimes. In addition, the need to have special charging stations is eliminated as well as the need for human involvement to plug in the charging cable. The objective of this paper is to report on methodological questions related to the potential usage of that charging technology for airport apron vehicles. We focus on the exemplary case of passenger buses transporting passengers from and to aircraft standing at outside parking positions. Still, we are convinced that the results hold for other types of service vehicles as well.

In order to charge apron vehicles with this technology, a dynamic inductive charging infrastructure would have to be implemented on the airport apron. This infrastructure consists of two components: the Power Supply Unit (PSU) and the Inductive Transmitter Unit (ITU). The PSU provides an alternating current of the required frequency to the ITU.

The ITU is installed below the road surface and charges the battery if a vehicle travels along (Panchal et al. [6]). Since the infrastructure requires high initial investments, only a fraction of the road network should be equipped with a charging track. At the same time, however, it must be spatially allocated in such a manner that the vehicles' batteries can be sufficiently charged while they are operating.

We use mathematical optimization models to formally characterize the problem of finding a spatial placement of the required infrastructure components such that the necessary capital investment is as small as possible. First models for planning inductive charging infrastructures on airport aprons have already been developed (Helber et al. [7] and Broihan et al. [8]). The standard approach to solving those models is to employ high-end commercial mixed-integer linear programming solvers such as Gurobi or CPLEX. However, Broihan et al. [8] have shown that solving real-world-sized test instances with standard solvers to proven mathematical optimality in a reasonable time is very often not possible.

This leads to the research questions addressed in this paper: Which features tend to make a particular instance of the infrastructure design problem of spatially allocating the components of the charging infrastructure on the airport apron road system hard to solve? Hard to solve in this context means that even within hours or days of computation time, it is not possible to find an infrastructure allocation that is known to be optimal in the mathematical sense of the underlying problem. If it turns out that this is indeed the case, a second question arises: Can we at least make a statement about the potential "optimality gap", i.e., indicating how far away from the optimal solution quality we can be at most? In order to answer these questions, we systematically generated a large-scale test bed of synthetic problem instances that reflect different types of real-world spatial structures of airport terminals as well as apron road networks and aircraft parking positions.

We will show that the proof of optimality takes a long time, although an admissible solution can already be found quickly. We will also investigate the influence of the problem's size and certain parameter specifications on the computation times. We show that the investments in the PSUs and ITUs, the vehicles' energy consumption and the energy intake can significantly impact the computation time.

To this end, we analyze the properties of the Dynamic Inductive Charging Problem (DICP) experimentally to determine why this problem is difficult to solve for standard solvers. In particular, we examine the problem properties that lead to high computation times. The structure of the paper is such that we first provide a brief overview of the inductive charging technology, characterize the resulting airport apron design problem and report on related literature in Section 2. In Section 3.1, we formulate the model assumptions based on the previously stated properties of airport aprons. The introduced model in Section 3.2 is a variant of the optimization model presented in Broihan et al. [8]. Section 4 describes the generation of our instances that we use in the numerical study. We describe the general instance generation process in Section 4.2 and characterize the properties of the generated instance set for the analysis in Section 4.3. The results of the numerical study are presented in Section 5. We analyze the properties of the different instances and relate them to the computation time. Section 6 summarizes the results of this paper.

2. Characterization of the Problem Setting and Related Research

2.1. Dynamic Inductive Charging

Dynamic inductive charging means that the vehicle is charged wirelessly while in motion. For the wireless power transfer, primary coils are installed below the surface at selected elements of the airport apron road system (see Figure 1). Such a so-called ITU is supplied with an alternating current of the desired frequency by the PSU, which in turn needs a connection to the power grid. The PSU uses power electronics to modify the frequency. SAEJ2954 defines the frequency for wireless power transfer for electric vehicles in the range of 81.39–90 kHz (SAE International [9]). The electromagnetic field is created locally below the pickup unit of the moving vehicle, say, a passenger bus. Via the secondary coils within the vehicle's pickup unit, an alternating current is induced by the electromagnetic field. This alternating current is rectified and used to charge the vehicle's battery and, eventually, power its electric engine. For further technical details on the technology of dynamic inductive charging and example projects, we refer to Li and Mi [10], Cirimele et al. [11], Lukic and Pantic [12], Panchal et al. [6], Ahmad et al. [13] and Covic and Boys [14].

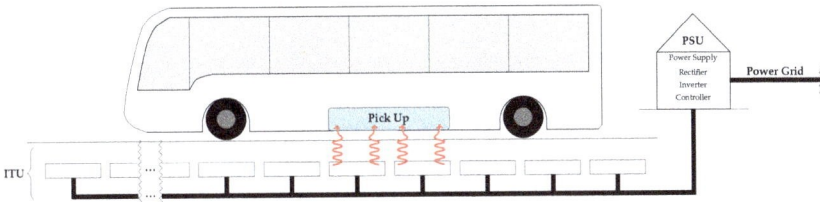

Figure 1. Components of the dynamic inductive charging system (Source: Broihan et al. [8]).

The efficiency of the system is strongly dependent on the gap between the primary and secondary coil (Imura and Hori [15] and Moon et al. [16]). The smaller the gap between the primary and secondary coils, the higher the power transmission efficiency. For this reason, dynamic inductive charging systems are suitable for flat surfaces on which low-profiled vehicles operate. This is exactly the situation on airport aprons and one of the reasons why this charging technology might be interesting for airport aprons.

2.2. Energy Density, Trip Structures and Modeling of Battery Levels

The energy density of electric batteries is known to be low relative to that of diesel fuel. Furthermore, the time required to transfer a certain amount of energy by charging its batteries, either conductively or inductively, is large compared to the time required to pump diesel fuel into the tank of a comparable vehicle with a combustion engine. Finally, an electric battery is not only expensive but also heavy due to its relatively low energy density. For all those reasons, the decision about the capacity of the battery of an electric vehicle is delicate from both the economic and the operational perspective: Very large batteries are not only expensive, but their transportation as part of the moving vehicle itself also consumes energy. On the other hand, small batteries require frequent re-charging and need to have ample spatially distributed charging facilities, again either for conductive or inductive charging.

For those reasons, many researchers studying dynamic inductive charging infrastructure design problems decided to model allocation decisions for ITUs together with battery size decisions for vehicles. The typical assumption is that a vehicle, say, a passenger bus serving an urban bus line, starts with a full battery at some initial location A, travels along a pre-defined route while serving a sequence of bus stations, and ends the tour at some final destination B. The charging infrastructure has to be allocated in such a way that the vehicle is never confronted with an empty battery while on its trip. To this end, the State of Charge (SOC) of the battery is tracked meticulously, considering both phases of de-charging and phases of charging (while passing ITU-equipped segments of the road system). An underlying assumption is that if the bus reaches the end of its tour, all that is needed is a battery that is not empty and that the battery will be fully charged before the bus begins its next trip. Examples of those modeling approaches can be found in Ko and Jang [17], Hwang et al. [18] and Ko et al. [19]. An important result of those studies is that the optimal structure of the charging infrastructure depends on the number of vehicles using it. Suppose only a small number of vehicles use the charging structure. In that case, it is beneficial to equip those few vehicles with large (and expensive) batteries to need only a few charging segments along the route the vehicles will travel. It is, however, not attractive to have a very large number of those vehicles equipped with large and expensive batteries. In this case, it is economically advisable to have a larger fraction of the road segments equipped with ITUs and to be able to operate with smaller (and, hence, less costly) battery sizes in many vehicles.

When considering the question of how to allocate the ITUs and PSUs within the road network of an airport apron, it turns out that the spatial structure as well the nature of the trips driven by, say, passenger buses, differs substantially from those found in urban mass transportation by public buses.

Figure 2 presents as an example a selected part of Vienna Airport. On the airport apron, a road network connects the aircraft parking positions, equipment service areas, vehicle depots and terminal buildings adjacent to the airport apron. The parking positions for aircraft can be distinguished between gate positions and outside positions (Mensen [20]). Passengers can reach an aircraft parked at a gate position via a boarding bridge, while apron buses are used for transportation to the outside parking positions.

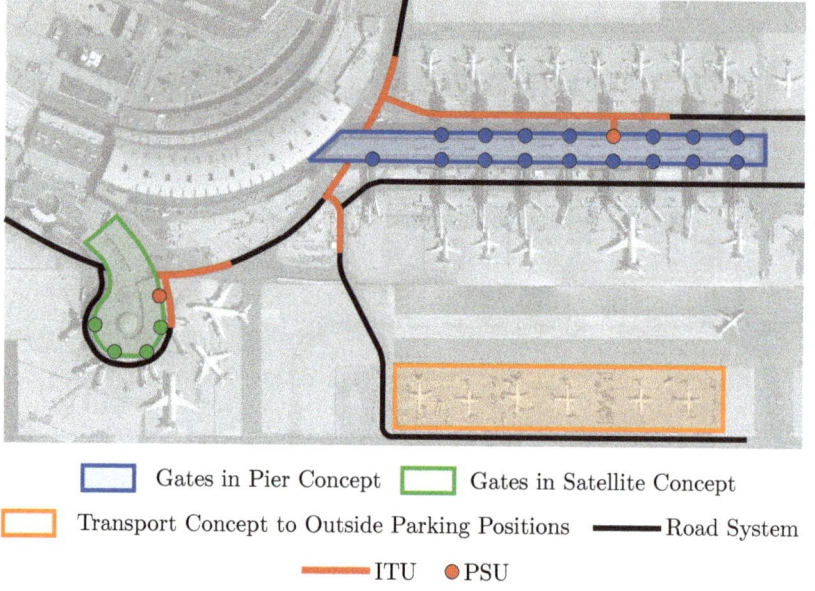

Figure 2. Selected part of the apron at Vienna Airport. Source: Vienna Airport [online], 48°07′03.39″ N 16°33′52.75″ E, Height 785 m, Google Earth © GeoBasis-DE/BKG 2009, URL: http://www.google.com/earth on 19 April 2022.

The layout of the terminal buildings determines the location of gate parking positions and the passenger gates. There are different terminal layout concepts, such as the pier, the satellite and the linear design. Figure 2 shows that these three layout concepts co-exist at Vienna airport.

At each terminal, several gates are available. Some gates are used exclusively to let passengers (un-)board their aircraft via a passenger bridge or to use passenger buses to transport the passengers to or from an outside aircraft parking position. In contrast, other gates can operate both with passenger bridges or passenger transportation buses.

During the turnaround of an aircraft, apron vehicles travel to and from the specific parking position, as well as gates or depots. A passenger bus, the exemplary type of vehicle considered in this paper, could pick up passengers at an aircraft and transport them to a terminal gate. Afterward, the bus could travel to another gate to pick up passengers and transport them from that gate to the parking position of their respective aircraft.

If we compare the operational elements of trips driven by passenger buses on airport aprons to those in urban public transport, we see three important differences:

- The distance and duration of a passenger bus trip on an airport apron are typically relatively short compared to those in public urban transportation.
- Most trips either start or end at one of the terminals or other dedicated airport areas.
- The number of different possible routes driven to serve the many possible service requests can be relatively large if we combine the different starting points, say, at the different terminal gates, with the many different destinations, for example at the outside aircraft parking positions.

Due to the third point, it seems impractical to follow the very detailed modeling approach introduced in Ko and Jang [17] and track at a very fine-grained resolution the SOC for the potentially extremely large number of conceivable service requests.

For this reason, we decided in this paper, as in Helber et al. [7] and Broihan et al. [8], to use a fundamentally different modeling approach. We require that the energy intake must be higher than the energy consumption for every service request. Therefore, the battery charge level at the end of a service request cannot be lower than at the beginning. As a result of that modeling decision, we do not need to model the SOC of the vehicles' battery in detail.

2.3. Modeling the Spatial Allocation of ITUs and PSUs

In Figure 2, we show in red a part of a fictitious allocation of ITUs and PSUs on some of the roads (in black) used by passenger buses for the Vienna Airport. Each ITU has to be part of a contiguous structure which is in turn connected to a PSU. Some locations, in particular those close to buildings, might be natural candidates to establish a PSU. On the other hand, there may be parts of the apron road network where installing ITUs could be impossible or unattractive due to interference with aircraft or the nature of road surfaces.

To represent the airport apron in the mathematical optimization model, we model the road network as a directed graph, as shown in Figure 3. Gates, parking positions and road intersections are modeled as nodes. The lanes of the airport apron roads are represented as links (directed arcs) in the graph. We assume that, on these links, the ITUs can be installed. If there are multiple adjacent edges, each with an installed ITU in the graph, they represent one contiguous ITU structure on the real airport apron. An example is given in Figure 3, where the ITU is installed between the nodes $g2$, $i6$, $i5$, $i14$ and $i15$. This contiguous ITU structure can then be powered by one PSU, which is installed at node $g2$ in the example. A consistent connection from each ITU segment to a PSU is required. This connection can be set up directly if the ITU is directly adjacent to a PSU node or indirectly via other ITU segments (e.g., the ITU segment between $i5$ and $i6$ is connected via the ITU segment between $g2$ and $i6$ to the PSU at node $g2$).

We are now in the position to state in a non-technical manner the infrastructure allocation problem. The overall objective is to minimize the investment in ITUs and PSUs. The selection of links to be equipped with ITUs as well as the installation of PSUs must be such that:

- Each link equipped with an ITU is either connected directly to a PSU or indirectly via a neighboring ITU-equipped link on the shortest route to their respective common PSU;
- For each service request, the vehicle can take up at least as much energy while driving along the relevant links as it needs.

This modeling approach relieves us from the need to track the SOC of the vehicles' batteries. Figure 4 illustrates this. It shows an example SOC curve for a service request. The service request consists of the links $l1$ to $l6$. At links $l2$ and $l5$, an ITU is installed. When the vehicle passes over an ITU, it absorbs energy and the curve increases. In all other cases, the SOC decreases. According to our assumption, the vehicle must absorb at least as much energy as it consumes for the service request. For this reason, the SOC at the end might be greater than at the beginning. Of course, it may happen that the vehicle cannot absorb the energy because the battery is already full. In that case, the SOC can be lower at the end. We assume the battery is large enough to survive a longer part of the service request without energy intake. This results in the SOC never falling below zero.

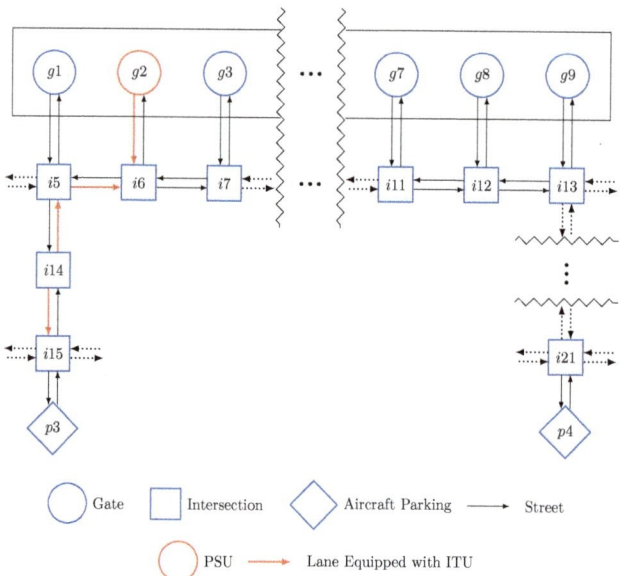

Figure 3. Representation of an airport as a directed graph. (Source: Adapted from Broihan et al. [8]).

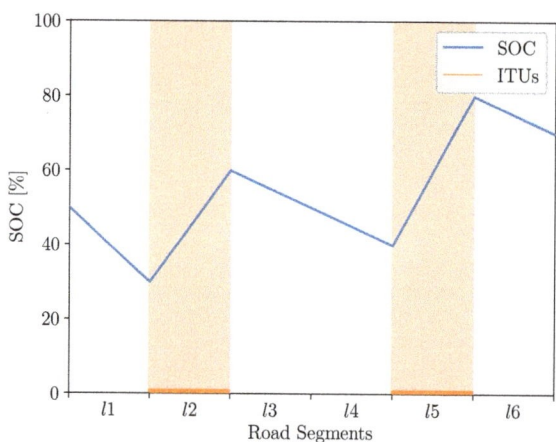

Figure 4. Exemplary SOC. (Source: Adapted from Broihan [21]).

2.4. Related Literature

For an overview of related literature on planning dynamic inductive charging infrastructure, we refer to Jang [22], Majhi et al. [23] and Yatnalkar and Narman [24]. They provide a comprehensive review of research articles and pilot projects. Most of these projects consider implementing a dynamic charging infrastructure for public bus systems. Many papers also consider battery capacity in the planning. Jang et al. [25] formulate a model in which the battery size is determined in addition to the placement of the charging infrastructure. The SOC is also considered in the model. The route is divided into segments that are either fully equipped with an ITU or not equipped at all. Ko and Jang [17], on the other hand, consider the route continuously. However, the model presented for this purpose is nonlinear. Both papers consider only one bus route at a time. In contrast, Hwang et al. [18] describe a multi-route environment. The model presented here has the special feature that the vehicles' capacities can be different. Liu and Song [26] consider stochastic elements in their model formulation for planning a charging infrastructure. They first present a deterministic model and then a model with uncertain travel time and energy consumption. The reason why particularly public buses are considered is that the tours are known in advance. The papers often consider so-called closed environments. That means systems where the influence of traffic and other external factors is small. In this case, the tour's energy consumption and intake can be easily determined. Some papers consider other environments, e.g., Schwerdfeger et al. [27] looks at planning a charging infrastructure for highways.

Only Helber et al. [7] and Broihan et al. [8] considered the optimal placement of wireless charging infrastructures on airport aprons. Helber et al. [7] studied the characteristics of dynamic inductive charging on airport aprons and introduced the first mathematical optimization model for planning such an infrastructure. Broihan et al. [8] presented a reformulation and extension of this model considering multiple vehicle types and a service level restriction. In a numerical study, they analyzed test instances based on real airport aprons. For most of the instances, they could not prove optimality within a time limit of seven days with a standard solver. The resulting optimality gaps ranged from 8% to 15%. However, with the analyzed instances, they could not identify the reasons behind high computation times and the problem properties that lead to those. It is exactly this open question that we address in this paper.

3. The Dynamic Inductive Charging Problem

3.1. Notation and Assumptions

To formally define the DICP, we now present the modeling assumptions as well as the notation summarized in Table 1. Note that assumptions and notation are closely related to those described in Helber et al. [7] and Broihan et al. [8], as the DICP is a generalization of the Dynamic Inductive Charging Problem considering Multiple Vehicle Types (DICP-MV) introduced in Broihan et al. [8].

Table 1. Notation used in the DICP model.

Indices and Sets	
$v \in \mathcal{V}$	vertices $\mathcal{V} := \{1, \ldots, V\}$
$l \in \mathcal{L}$	links $\mathcal{L} := \{1, \ldots, L\}$
$r \in \mathcal{R}$	service requests $\mathcal{R} := \{1, \ldots, R\}$
$\mathcal{P} \subseteq \mathcal{V}$	set of vertices $v \in \mathcal{V}$ qualified to host a PSU
$\mathcal{P}_l \subseteq \mathcal{V}$	set of PSU candidates able to supply link $l \in \mathcal{L}$
$\mathcal{L}_v \subset \mathcal{L}$	set of links $l \in \mathcal{L}$ qualified to be powered by a PSU candidate at $v \in \mathcal{P}$
$\mathcal{L}_r \subseteq \mathcal{L}$	set of links $l \in \mathcal{L}$ included in service request $r \in \mathcal{R}$
$\mathcal{LP}_v \subset \mathcal{L}$	set of links $l \in \mathcal{L}$ directly neighboring a PSU candidate at $v \in \mathcal{P}$
$\Gamma_{lv} \subset \mathcal{L}$	set of predecessors $l' \in \mathcal{L}$ of link $l \in \mathcal{L}$ in a direct connection on the shortest path to a PSU candidate at $v \in \mathcal{P}$
Parameters	
$c_v^{psu} \in \mathbb{R}^{\geq 0}$	investment in a PSU at vertex $v \in \mathcal{P}$
$c_l^{itu} \in \mathbb{R}^{\geq 0}$	investment in an ITU at link $l \in \mathcal{L}$
$ei_l \in \mathbb{R}^{\geq 0}$	energy intake by traveling on link $l \in \mathcal{L}$
$ec_r \in \mathbb{R}^{\geq 0}$	energy consumption by serving request $r \in \mathcal{R}$
Decision Variables	
$X_{lv} \in \{0, 1\}$	1, if link $l \in \mathcal{L}$ is equipped with an ITU and powered by PSU at vertex $v \in \mathcal{P}$, 0 else
$Y_v \in \{0, 1\}$	1, if a PSU is installed at vertex $v \in \mathcal{P}$, 0 else

- We model the road system of the airport apron as a directed and weighted planar connected digraph $G(\mathcal{V}, \mathcal{L})$ composed of a set of vertices $v \in \mathcal{V}$ and a set of directed edges or links $l \in \mathcal{L}$ (see Figure 3).
- A vertex $v \in \mathcal{V}$ defines a particular geographic location on the airport apron. This includes, for example, gate positions, aircraft parking positions, depots or fixed positions in the road system to split road sections.
- We define a subset $\mathcal{P} \subseteq \mathcal{V}$, which denotes the set of vertices $v \in \mathcal{V}$ qualified to host a PSU. Installing a PSU at a candidate vertex $v \in \mathcal{P}$ requires a fixed investment $c_v^{psu} \in \mathbb{R}^{\geq 0}$.
- Each directed link $l \in \mathcal{L}$ represents a lane segment of the airport's road system. It connects a pair of adjacent vertices $v \in \mathcal{V}$. The direction of each link $l \in \mathcal{L}$ corresponds to the direction of travel on this lane segment.
- We assume every link $l \in \mathcal{L}$ to be a candidate to host an ITU. To supply such an ITU with electricity, it must be connected to a PSU, either directly or indirectly, via some other link. We define a set of links $\mathcal{L}_v \subset \mathcal{L}$ which are qualified to be powered from a PSU installed at vertex $v \in \mathcal{P}$.
- Similarly, we define $\mathcal{P}_l \subseteq \mathcal{V}$ as the set of vertices that could be equipped with a PSU powering link l.
- To ensure a physical connection from each ITU to its power-supplying PSU, we define a set of links $l' \in \Gamma_{lv} \subset \mathcal{L}$ for each combination of link l and PSU candidate v as $\Gamma_{lv} = \{l' \in \mathcal{L}_v \mid l' \text{ precedes } l \text{ on a shortest path to } v \in \mathcal{P}\}$.
- $\mathcal{LP}_v \subset \mathcal{L}$ defines a set of links l directly neighboring a PSU candidate $v \in \mathcal{P}$.
- The installation of an ITU at link $l \in \mathcal{L}$ requires a fixed investment denoted by c_l^{itu}.
- We define service requests $r \in \mathcal{R}$ that represent the apron vehicles' potential service tasks, e.g., passenger transfers from a gate to an aircraft or baggage transportation.
- Serving request $r \in \mathcal{R}$ requires the vehicle to move along a set of links $l \in \mathcal{L}_r$.
- We denote the consumed energy for a particular request $r \in \mathcal{R}$ by traveling along the links in $\mathcal{L}_r \subseteq \mathcal{L}$ by ec_r.
- As the vehicle serving request r travels along an ITU-equipped link $\mathcal{L}_r \subseteq \mathcal{L}$, it can take up to ei_l units of energy and charge its battery.
- For each request r, the sum of the potential energy intakes ei_l by traveling along the links in $\mathcal{L}_r \subseteq \mathcal{L}$ must be at least as large as the energy consumption ec_r for that request. This assumption relieves us from the need to model the battery's SOC.

To describe the arrangement of charging infrastructure across the airport apron, we introduce two binary decision variables. The variable $Y_v \in \{0, 1\}$ takes a value of 1 if a PSU is installed at the node $v \in \mathcal{P}$ and 0 otherwise. Likewise, $X_{lv} \in \{0, 1\}$ equals 1 if an ITU is installed at link $l \in \mathcal{L}_v$ and connected to a PSU at vertex $v \in \mathcal{P}$ and 0 otherwise.

3.2. Model Description

Based on the previous assumptions, we introduce the DICP, which is a generalization of the DICP-MV presented in Broihan et al. [8], as a linear program in binary variables as follows:

$$\min F = \sum_{v \in \mathcal{P}} \left(c_v^{psu} \cdot Y_v + \sum_{l \in \mathcal{L}_v} c_l^{itu} \cdot X_{lv} \right) \tag{1}$$

such that

$$\sum_{v \in \mathcal{P}} \sum_{l \in \mathcal{L}_r \cap \mathcal{L}_v} X_{lv} \cdot ei_l \geq ec_r \qquad r \in \mathcal{R} \quad (2)$$

$$X_{lv} \leq Y_v \qquad v \in \mathcal{P}, l \in \mathcal{L}_v \quad (3)$$

$$\sum_{v \in \mathcal{P}_l} X_{lv} \leq 1 \qquad l \in \mathcal{L} \quad (4)$$

$$\sum_{l' \in \Gamma_{lv}} X_{l'v} \geq X_{lv} \qquad v \in \mathcal{P}, l \in \mathcal{L}_v \setminus \mathcal{LP}_v \quad (5)$$

$$X_{lv}, Y_v \in \{0,1\} \qquad v \in \mathcal{P}, l \in \mathcal{L}_v \quad (6)$$

The objective function (1) minimizes the total investment in the components of the dynamic-inductive charging infrastructure. Constraint (2) ensures a sufficiently large energy intake while serving a request $r \in \mathcal{R}$. According to constraint (3), ITU and PSU installation decisions are connected. Restriction (4) ensures that each link can be equipped at most once with an ITU, in which case it is connected to exactly one installed PSU. Restriction (5) enforces a connected ITU infrastructure from the furthest ITU along a shortest path to the powering PSU.

4. Generating a Set of Instances for Dynamic Charging Infrastructures Problems
4.1. Purpose and Objective of the Instance Generation Process

Our previous and preliminary numerical results (see Helber et al. [7] and Broihan et al. [8]) showed that any attempt to use a high-end commercial mixed-integer programming solver like Gurobi or CPLEX to solve the DICP shows very mixed results with respect to computation times and solution quality. In particular, we observed that when the instances

- tended to have a relatively small number of links l and vertices v in the graph representing the road network on the airport's apron and, hence, also
- tended to have a relatively small set of service requests r to be considered,

is then those commercial solvers could often solve the resulting instances of the DICP to proven optimality within a few seconds or minutes.

However, real-world airports often have large and complex apron road networks, many passenger gates and many aircraft parking positions. As one consequence, the operational variance of the possible routings of the vehicles (passenger buses in our example) can be substantial. As we aim at obtaining a charging infrastructure allocation that is robust over a wide variety of such service requests, many of them have to be considered simultaneously in the infrastructure design decision, which is one factor leading to large model instances that tend to be hard to solve, i.e., having intolerably long computation times as well as potentially large optimality gaps.

A further problem of dealing with larger real-world airports is that the length of road segments, say, between terminals and aircraft parking positions, can be substantial. It could be desirable to equip only small fractions of those long road segments with the ITUs. However, to represent those fractions of the road segments in our model, we have to subdivide those road segments into sub-segments by adding additional vertices to the graph. An example of this problem aspect is depicted in Figure 5. Here, between the two nodes denoted as $i44$ and $i51$, two long road segments exist, one for each direction and each having a length of 300 m. By introducing two or even five further vertices, link lengths of 100 m or even 50 m are created, respectively.

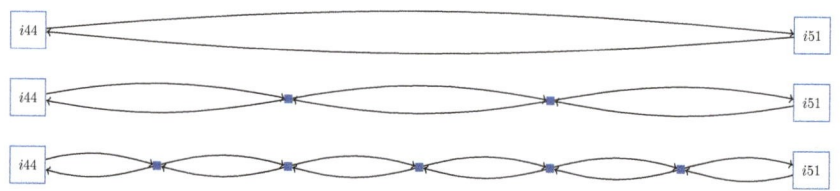

Figure 5. Extending an initial graph with link length maximum (**top**), 100 m (**mid**) and 50 m (**bottom**).

Having such finer granularity of the road system network in our model makes more cost-efficient infrastructure solutions possible, which is attractive. However, this comes at the price of operating again with larger models, i.e., models with a larger number of links and vertices that tend to be numerically more difficult to solve.

Another element that affects the difficulty of solving the problem by using standard commercial solvers to proven optimality is the cost ratio of the elements of the charging infrastructure, i.e., the necessary investment per PSU relative to the investment per ITU unit. If the PSUs tend to be relatively expensive, the resulting structures tend to operate with smaller numbers of PSUs to which then relatively large ITU structures are connected. In the opposite case of relatively inexpensive PSUs, larger numbers of ITU structures are placed over the road network.

Finally, the ITUs' power transfer also significantly impacts both the structure of the solutions and the difficulties of finding them. Suppose the ratio of the energy that can be picked up by a vehicle passing along a link is very large relative to the energy required to pass along that link. In that case, it may be possible to equip only a relatively small but well-chosen fraction of the apron road network with ITUs.

We know that all these factors affect:

- The spatial structure of solutions to the infrastructure design problems;
- The computational time to find those solutions when using a standard solver such as Gurobi.

We also conjecture that there might be cross-effects between the influencing factors. In order to be able to identify those effects and to explore the limitations of using a commercial solver to solve the DICP, we systematically designed a full-factorial test bed consisting of hundreds of instances. We then solved those instances numerically to obtain experimental results shedding some light on the questions outlined above. Before we present the results of those computations, we first describe the underlying system of creating the test instances as well as their characteristic features.

4.2. Design of the Instance Generation Process

For our numerical study, we need to be able to create many instances systematically. Therefore, we developed an instance generator that was implemented in Python 3.8. The procedure to generate instances can be divided into the three main steps of (i) creating a so-called initial graph, (ii) deriving from this initial graph a so-called instance graph, and (iii) adding further sets and parameters to arrive at a complete description of a planning instance.

In our case, the initial graph is created using the Python graph modeling package NetworkX and based on real airport apron structures. With the help of a satellite image, as presented in Figure 6 for a small part of Tokyo Airport Haneda, Japan, the graph nodes can be set according to the real airport's intersections, gates and parking positions. The weights of the edges correspond to the road segment's lengths of the real airport. An example of a complete initial graph is given in Figure 7; Hamburg Airport, Germany, inspires this one.

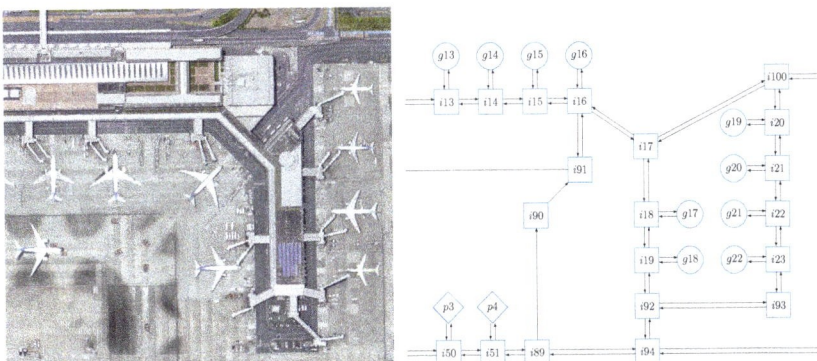

Figure 6. Comparison between excerpt of satellite image and excerpt of generated initial graph. Source: Tokyo Airport Haneda [online], 35°33′16.41″ N 139°47′17.42″ E, Height 6 m, Google Earth ©, URL: http://www.google.com/earth on 11 July 2022.

In the second step, we derived from the initial graph different instance graphs classes by deleting and adding nodes and edges to create instances that differ concerning the granularity of the modeled topology, resulting in instance classes denoted as "small", "medium" and "large". Graphs of the small instance class have only a small number of nodes and links. Thus, we must delete nodes and edges of the initial graph to adapt it to the small instance class. For the large instance class, we need to add nodes and edges to the initial graph. Within an instance class, the graphs derived from different initial graphs should have a comparable number of nodes and links. Since the initial graphs have different sizes, the number of links and nodes to be deleted or added differs for each initial graph. In the following, we will explain the mechanisms for deleting and adding nodes to create comparable instance graphs based on very different initial graphs.

Graph reduction is done by deleting gates, depots and parking positions and their associated intersections. For this purpose, the user specifies the portion of these positions to be deleted (deletion rate). The positions are then deleted at regular distances. Figure 8 shows an excerpt from the initial graph introduced in Figure 7 for different deletion rates. The figure at the top shows the result of a deletion rate of 0% (i.e., the initial graph), in the middle that of a deletion rate of 50%, and at the bottom the result of a deletion rate of 75%.

In addition to deleting nodes and edges, it is also possible and may be necessary to add them as explained in Section 4.1 and Figure 5 to achieve a finer granularity of the modeled topology and, hence, economically more efficient solutions.

In the third step, we generate the final instances for each instance graph. To this end, different parameters and sets are required, as indicated by the notation in Table 1. Table 2 summarizes how we derived the sets and parameters of the DICP. The set of links $\mathcal{L}$ and nodes $\mathcal{V}$ are directly taken from the instance graph. We consider all links to be candidates to host an ITU. Limiting the consideration to heavily trafficked sections of the road network would be possible. However, in this case, there is a risk that not all service requests can be served. This is particularly true if service requests do not use these route sections. For this reason, we consider all links as ITU candidates. We determine the PSU candidate positions $\mathcal{P}$ from the set of all depots and gate positions to achieve a predetermined number of candidate positions so that those positions are evenly spread over the set of depots and gate positions. We assume that each PSU can supply each link. For this reason, $\mathcal{L}_v$ equals $\mathcal{L}$ for all PSU candidates. Conversely, this also means that every PSU can supply every link and, thus, $\mathcal{P}_l$ equals $\mathcal{P}$ for all links. Again, this is a simplified assumption. In reality, it would be conceivable that PSUs could not supply ITUs at any distance. This could be realized in our model if we consider in the set $\mathcal{L}_v$ only the ITUs within a certain distance from the PSU v.

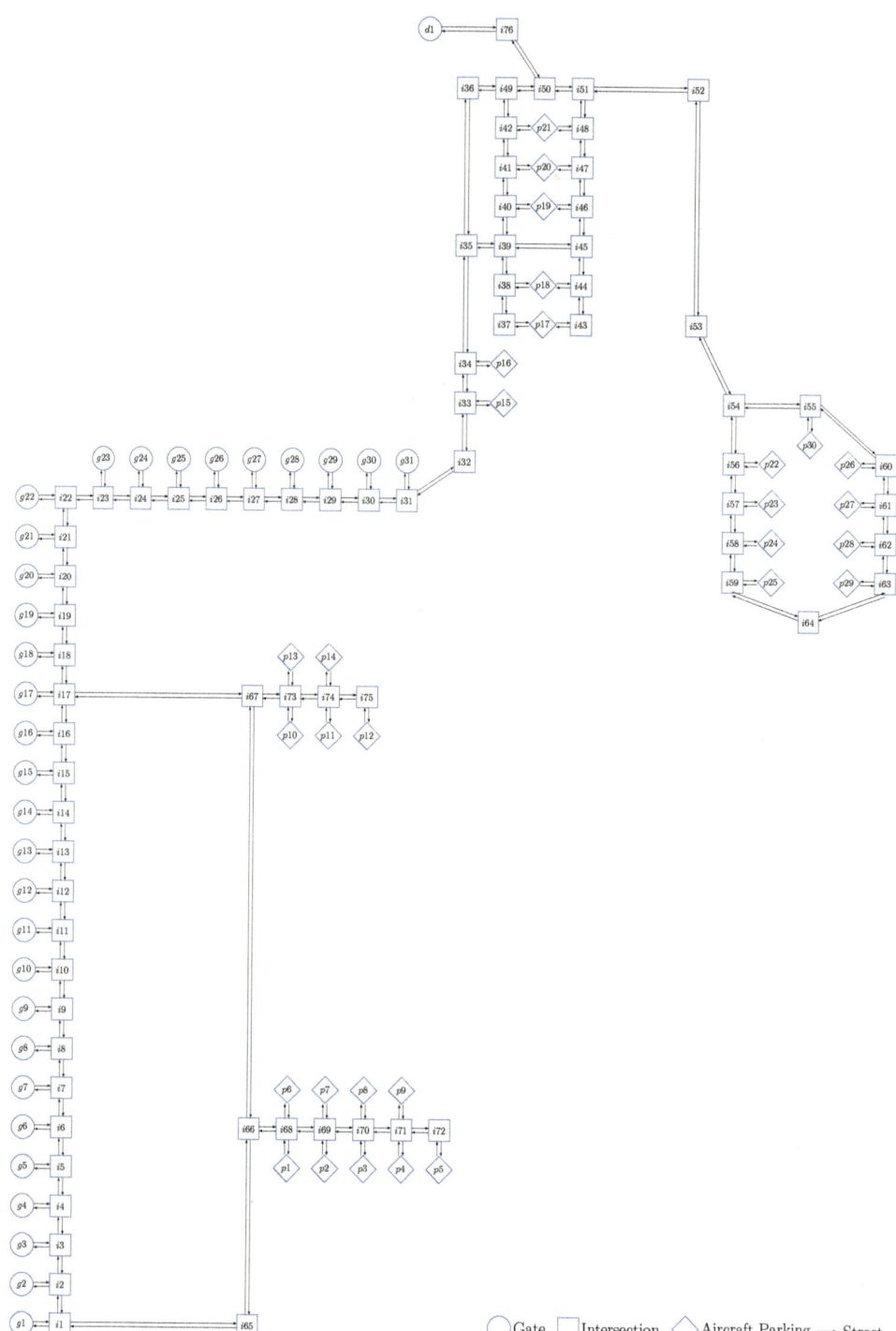

Figure 7. Example of an initial graph (Hamburg Airport).

Figure 8. Reducing an initial graph with deletion rate 0% (**top**), 50% (**mid**) and 75% (**bottom**).

Modeling the service requests is a crucial aspect of setting up problem instances. In all instances, we use the two-way request structure described in Broihan et al. [8]. Thus, a request starts at a terminal position (gate or depot), takes the shortest path to an aircraft parking position and ends again at a terminal position. All combinations of starting points (terminal positions), aircraft parking positions and endpoints (terminal positions) form the complete set of possible requests. Consequently, even in medium-sized instances, we get a very large set of requests. For the case of the initial graph in Figure 7, we have 31 passenger gates plus one bus depot and 30 outside aircraft parking positions. As a result, we have $(31+1) \times 30 \times (31+1) = 30{,}720$ different two-way requests, leading to 30,720 constraints (2) in the DICP (1)–(6). For other (larger) airports the number of requests and corresponding model constraints may be substantially larger. However, these service requests often overlap, so it can be sufficient to consider only a part of them. Hence, we operate with different proportions of all possible requests. According to the given proportion, the requests for the set $\mathcal{R}$ to be considered in the model are randomly selected from all possible requests for the given instance graph. We assume that the vehicles take the shortest path for each request and, hence, used shortest path algorithms to determine the set $\mathcal{L}_r$ of links over which a vehicle travels as it serves request r. In addition, a set Γ_{lv} represents the predecessor of a link l on the shortest path to a PSU p. This set is determined using the Python package NetworkX in the preprocessing step to determine the shortest paths for all the requests.

Table 2. Instance derivation from given data.

Input	Graph, energy intake per meter, energy consumption per meter, investment per PSU, investment in ITU per meter, percentage of requests, number of PSU candidates
Output:	$\mathcal{L}, \mathcal{V}, \mathcal{P}, \mathcal{L}_v, \mathcal{P}_l, \mathcal{R}, \mathcal{L}_r, \Gamma_{lv}, c_v^{psu}, c_l^{itu}, ei_l, ec_r$
Sets:	
$\mathcal{L}$	given by instance graph
$\mathcal{V}$	given by instance graph
$\mathcal{P}$	given number of PSU candidates is selected from graph
$\mathcal{L}_v$	equals $\mathcal{L}$ for all $v \in \mathcal{P}$
$\mathcal{P}_l$	equals $\mathcal{P}$ for all $l \in \mathcal{L}$
$\mathcal{R}$	from all given two-way request structures a given percentage of requests is chosen
$\mathcal{L}_r$	derived from graph
Γ_{lv}	derived from graph
Parameters:	
c_v^{psu}	equals given investment per PSU
c_l^{itu}	link length (derived from graph) is multiplied with given investment in ITU per meter
ei_l	link length (derived from graph) is multiplied with given energy intake per meter
ec_r	request length (derived from graph) is multiplied with given energy consumption per meter

We assume that the investment c_v^{psu} in a PSU at node v is equal for all $v \in \mathcal{P}$ and is provided by the user. The investment c_l^{itu} to equip link l with an ITU is derived by multiplying the investment in ITU per meter with the link length. Similarly, the energy intake ei_l is determined by multiplying the energy intake per meter with the link length and the energy consumption ec_r by multiplying the energy consumption per meter with the request length. The length of the request is determined by adding the lengths of all links contained in this request.

4.3. Description of the Generated Instances

For the numerical study conducted in this paper, we created a test instance set based on three initial graphs. These graphs are based on sections of real airport aprons but underwent minor modifications. Each graph in Figure 9 represents a certain apron layout. We refer to them as structures A, B and C. General layout A (left part of Figure 9) is an aggregate representation inspired by the conditions found at Hamburg airport, for which the initial graph was already introduced in Figure 7. The other two structures B and C (center and right part of Figure 9) were inspired by topologies found at other international airports, again without actually being isomorphic representations. Note that in these aggregate visualizations of Figure 9, an entry such as "G1" denotes an entire set of terminal gates in close proximity. Likewise, an entry "P1" represents an entire group of aircraft parking positions.

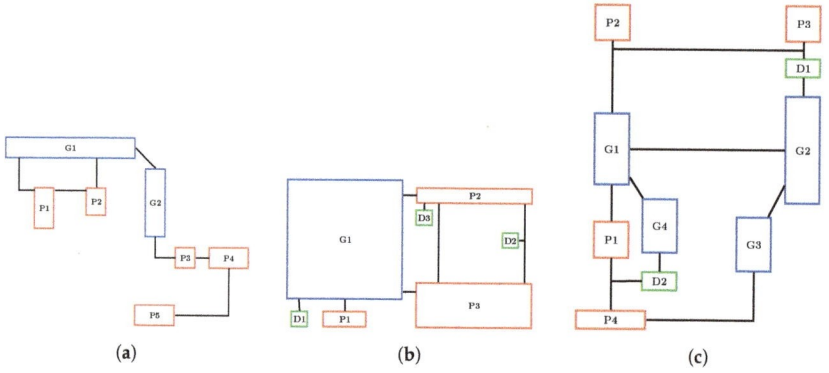

Figure 9. Aggregate Layout Structures. (**a**) Structure A. (**b**) Structure B. (**c**) Structure C.

We derived the initial graphs from satellite images. We set the positions for parking positions, terminal positions and intersections according to these images. Additionally, we placed intersection nodes at positions relevant to the structure (e.g., curves).

As mentioned before, we defined three instance classes of different sizes. Graphs of the instance class "small" have 55 to 65 nodes, graphs of the instance class "medium" have 100 to 120 nodes, and graphs of the instance class "large" have 180 to 200 nodes. We used different vertex (node) deletion rates and resulting link lengths to generate these instance graphs from each initial graph for structures A, B and C. The link length always specifies a maximum length since the links of the initial graph cannot always be divided without a remainder. Suppose we consider an instance graph with a link length of 100. In this case, a 250 m link of the initial graph will be divided into two links with a length of 100 m and one with a length of 50 m.

The values of deletion rates and link length are shown in Table 3. Excluding the initial graphs, this led to three different graphs per instance class and, hence, a total of nine different instance graphs.

Table 3. Graph Adjustments.

Instance Class	Structure	Nodes	Deletion Rate	Link Length
Base	A	138	0%	max.
	B	191	0%	max.
	C	293	0%	max.
Small	A	56	75%	max.
	B	57	85%	max.
	C	63	85%	max.
Medium	A	105	45%	100 m
	B	114	65%	200 m
	C	111	70%	200 m
Large	A	189	0%	50 m
	B	192	30%	100 m
	C	194	40%	100 m

In the numerical analysis, we analyzed the influence of selected sets and parameters on the computation time. Therefore, the following further attributes were varied when creating the instances:

- $\%|\mathcal{R}|$: Proportion of considered service requests out of those potentially possible for the given instance graph;
- $|\mathcal{P}|$: Number of PSU candidates;
- c^{psu}/c^{itu}: Ratio of investment per PSU and investment in a meter of an ITU;
- ei/ec: Ratio of energy intake per meter and energy consumption per meter.

By modifying the size of the set $|\mathcal{P}|$ of nodes potentially hosting a PSU and the size $\%|\mathcal{R}|$ of the set of service requests to consider, we varied the size of the problem. Therefore, we expected the computation time to increase for larger values for both attributes. However, we wanted to examine how significant the increase is. For the investment and energy parameters, only their ratios are relevant. As already explained, a high c^{psu}/c^{itu} ratio means that PSUs are expensive relative to ITUs. Consequently, few PSUs are expected to be built. Instead, more ITUs are built to connect all ITUs to the few installed PSUs. Thus, some ITU segments are built not because of the energy requirements of the vehicles but to produce a permissible (connected) charging infrastructure. A low c^{psu}/c^{itu} ratio can, on the contrary, lead to significantly more PSUs and fewer ITUs. With an ei/ec ratio close to one, energy intake and consumption are almost identical. As a result, nearly the entire infrastructure must be equipped with an ITU. An energy ratio ei/ec significantly larger than 1 indicates that the energy intake is significantly greater than the energy consumption. In such a case, only relatively few ITUs will be built. The influence of the parameters on the resulting infrastructure seems to be very clear. However, the influence on the computation time is not obvious, which is why we considered them in our analysis.

We considered three different values for every attribute, as shown in Table 4. The values were chosen to cover a wide range so that the influence of the attributes should be visible. For example, for the energy ratio ei/ec, a very low value (1.2) was selected, a value that may correspond to reality (3.24) and a particularly high value (10). We considered all different attribute combinations for all graphs. We have four attributes with three different values and thus have $3^4 = 81$ instances for each graph of an instance class. Since we considered nine graphs, we obtained a total number of 729 instances. We solved these instances with Python 3.8 and Gurobi 9.1.0. Considering a strategic planning problem, we used a time limit for the computation of 48 h for each of the 81 × 9 = 729 problem instances of our numerical test bed. In other words, we allowed up to about four years (!) of total Central Processing Unit (CPU) time for the entire study. All computations ran on the Dumbo subcluster of Leibniz University Hannover compute facilities, which uses an Intel

Xeon E5-4650v2 @2.40 GHz processor. Clearly, without such a cluster, the computations would not have been possible.

Table 4. Attributes of generated instances.

Attribute	Values
$|\mathcal{P}|$	1, 4, 8
$\%|\mathcal{R}|$	30%, 70%, 100%
c^{psu}/c^{itu}	1, 25, 100
ei/ec	1.2, 3.24, 10

5. Numerical Results

5.1. Overview of Computation Times

Table 5 shows the average computation times ($\bar{t}$), the median value (t^{median}), the minimum computation time (t^{min}) and the maximum computation time (t^{max}) for each instance graph. Each row of the table gives the aggregate results over 81 different instances stemming from the systematic combination of the parameter entries in Table 4.

As expected, the computing time increases for larger instance classes, up to the maximum of 48 h allowed for the Gurobi solution process. Thus, the size of the network has a significant influence on the computation time. However, even within an instance class, the computing times vary remarkably. Structure C has significantly higher computing times in all instance classes, although the graph size is similar to the other structures. In addition, even within a instance graph, the computing times fluctuate enormously, as indicated by the wide range between minimum and maximum values.

Further, we observe that the median value is significantly smaller than the average for the small and medium instance classes. This is because, within an instance graph, many instances can be solved to proven optimality within seconds, minutes or a few hours, and some instances could not be solved in the given time limit (48 h). The outliers lead to high median values for these instance classes.

However, for large instances, the mean value is always smaller than the median value. This is because most instances could not be solved to optimality within the time limit and consequently have a reported computation time of 48 h.

Table 5. Computation Times.

		$\bar{t}$	t^{median}	t^{min}	t^{max}
small	A	142 s	14 s	1 s	2129 s
	B	204 s	57 s	1 s	4357 s
	C	719 s	155 s	1 s	10,322 s
medium	A	15 h	3 h	0 h	48 h
	B	10 h	1 h	0 h	48 h
	C	23 h	19 h	0 h	48 h
large	A	28 h	48 h	0 h	48 h
	B	34 h	48 h	0 h	48 h
	C	37 h	48 h	1 h	48 h

5.2. Analysis of Solution Process

Moreover, we studied the time needed by Gurobi to find the first feasible solution ($\bar{t}^{first}$), the time after which the solution did not improve anymore ($\bar{t}^{last}$), the size of the Linear Program (LP)-Gap ($\overline{Gap}^{LP}$), and the size of the final Gap ($\overline{Gap}^{final}$) when the time limit was reached. We calculated the LP-Gap as follows: $Gap^{LP} = (OFV^{opt} - OFV^{rel})/OFV^{rel}$, where OFV^{opt} indicates the objective function value of the optimal solution and OFV^{rel} indicates the LP relaxation solution's objective function value. With this value, we want to describe the quality of the LP relaxation. A low LP-Gap indicates a good LP relaxation and a high value indicates a bad LP relaxation.

Since we have a time limit of 48 h, optimality cannot be proven for all instances. Therefore, the final gap ($\overline{Gap}^{final}$) indicates the gap between the upper and lower bound at the end of the time limit.

In addition, we performed a second numerical investigation in which the average number of equally optimal solutions ($\bar{N}^{opt}$) was determined for the small instances. We consider solutions as equally optimal if they have the same objective function value as the optimal solution but differ in structure. Gurobi searches for all solutions with the same objective function value as the optimal solution and counts the number. Gurobi stops when all solutions are found or if the total number of equally optimal solutions is higher than 100,000. Due to the time limit, such a computation is not possible for the instances in the medium and large instance classes. Because of the upper limit, the average value of the equally optimal solutions $\bar{N}^{opt}$ is underestimated.

For all described metrics, the average values of 81 instances for each instance graph are shown in Table 6.

Table 6. Analysis of solution process.

		$\bar{t}$	$\bar{t}^{first}$	$\bar{t}^{last}$	$\overline{Gap}^{LP}$	$\overline{Gap}^{final}$	$\bar{N}^{opt}$
small	A	142 s	0 s	46 s	47%	0%	6251
	B	204 s	0 s	56 s	44%	0%	23,315
	C	719 s	0 s	256 s	56%	0%	26,525
medium	A	15 h	0 s	2 h	47%	0%	N/A
	B	10 h	0 s	1 h	57%	0%	N/A
	C	23 h	0 s	6 h	51%	1%	N/A
large	A	28 h	1 s	3 h	45%	4%	N/A
	B	34 h	0 s	9 h	48%	5%	N/A
	C	37 h	1 s	12 h	46%	7%	N/A

A first feasible solution is found on average in less than one second in each instance class. This shows that the problem's difficulty does not lie in finding a feasible solution. This is not surprising as a feasible (but expensive) solution can always be constructed by equipping each link with an ITU and then installing just one PSU.

Compared to the total computation time, a solution that is no longer improved is found after a relatively short time, particularly for small and medium-sized instances. Even for the hard-so-solve large instances, in most of the computation time, no improvement of the incumbent solution occurs. Figure 10 shows the upper and lower bound course stemming from the Branch-and-Bound process for one exemplary instance of a large-sized instance of structure A. We observe that a feasible solution is found very quickly. The upper bound, i.e., the objective function value of the best solution found so far, improves very strongly at first, then only very slightly. After about 3 h, the upper bound no longer improves. The lower bound slowly approaches the upper bound in the remaining ten hours until the optimization terminates with the proof of optimality of the Branch-and-Bound procedure used by Gurobi.

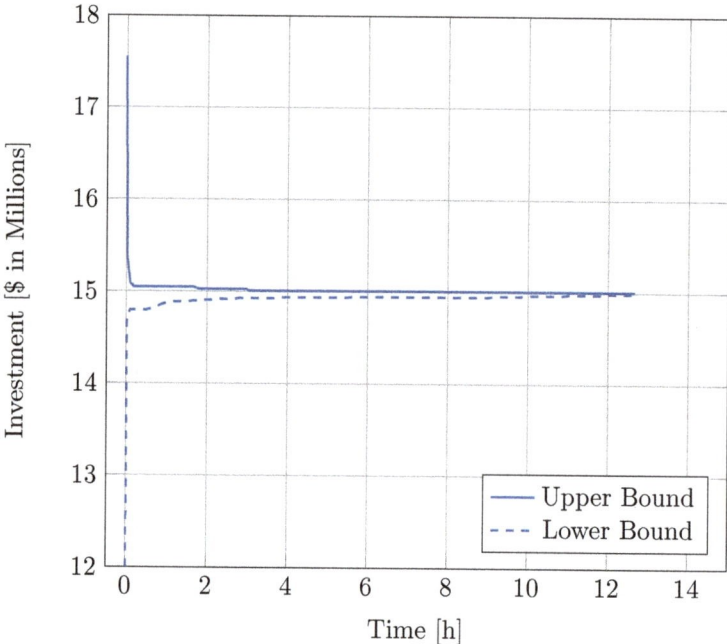

Figure 10. Solution Process of one Instance.

Table 6 shows that there are many equally optimal solutions for the instances of the small instance class. The high number of equally optimal solutions can be a reason for the long time needed to prove optimality. A large number of equally optimal solutions leads to many nodes to be visited in the Branch-and-Bound process, even if one optimal solution may already have been found. One reason for the high number of equivalent optimal solutions lies in the design of the instances. The links often have the same length and thus the same energy contributions and investments. Hence, many different infrastructure designs lead to the same energy contributions and investments. The structure also highly influences the number of equally optimal solutions and the computation time. If a structure is symmetrically built, many different infrastructures lead to the same energy contributions and investments. This is especially the case for structure C in our study. We observe the highest number of equally optimal solutions for this structure in the small instance class. But above all, we observe the highest computation times in all instance classes.

In addition, and making things worse, symmetric solutions occur, especially when the link length is reduced. Figure 11 shows an example of a symmetric solution. There is an intermediate node between nodes $i1$ and $i2$. Let all links have the same length. In this case, both represented solutions are equivalent because they contribute the same energy to each service request and have the same investment. In any route segment that is not interrupted by an intersection and in which the link lengths are the same, each solution is equivalent in that the sum of the ITUs built in either direction is identical. In the example, the sum of ITUs in each direction equals 1. Consequently, with smaller link lengths, more symmetries are expected since there are significantly more such route sections.

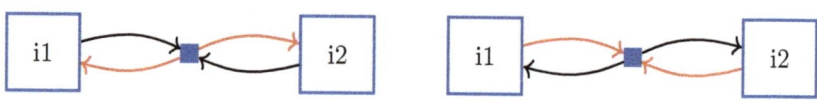

Figure 11. Example of a symmetric solution.

As expected, the final gap increases with increasing problem difficulty. Again, we observe that instances of structure C are more difficult to solve than those of the other two structures. Larger instances mean that it is more likely that the time limit will be reached. For this reason, the final gap is larger with increasing problem size.

The mean value of the LP-Gap has similar sizes for all structures (between 40% and 60%). However, the LP-Gap varies extremely between the individual instances. As Table 7 shows, the most influential factor is the energy ratio. The table shows the average values of the LP-gap for all instances together with the respective energy ratio $\frac{ei}{ec}$. For an energy ratio of 1.2, the average LP-Gap is only 5%, for a ratio of 3.24 it increases to 26% and for a ratio of 10 even to 115%. For a high energy ratio, it is sufficient to only partially equip links with an ITU, especially if the links are very long. In the solution of the LP-relaxation of the DICP, these links are "utilized" with very small fractionals such as 0.1. Therefore, the LP-Gap is very poor (far away from the ideal of 0%). However, if the energy ratio is low, many ITUs must be built, and partial equipment is not useful. Thus the LP-Gap is smaller, reducing the numerical effort of the Branch-and-Bound process performed by Gurobi. Unfortunately, this numerically attractive situation is not very interesting from a practical point of view. It corresponds to a situation in which ITUs would need to be installed almost everywhere, which is exactly what one would like to avoid in the attempt to find an economically efficient dynamic charging infrastructure.

Table 7. Influence of energy ratio on the LP-Gap.

		$\overline{Gap}^{LP}$
	1.2	5%
ei/ec	3.24	26%
	10	115%

5.3. Influence of Problem Properties on the Computation Time

The influence of the different instance attributes from Table 4 on the computation time is shown in Table 8. The computation times are divided according to structure and the attributes and averaged over 27 values. In the first cell, for example, the times are averaged over all instances that belong to the small instance class of structure A and have only one PSU candidate. They differ in the number of service requests, the energy ratio and the ratio of investments ($3 \times 3 \times 3 = 27$). In addition, the computation times of the medium-sized instances are presented graphically in Figures 12–15.

Table 8. Average computation times for different attributes.

		$|P|$			$\%|R|$			ei/ec			c^{psu}/c^{itu}		
		1	4	8	30%	70%	100%	1.2	3.24	10	1	25	100
small (in min)	A	0	1	6	1	2	4	5	2	0	4	2	1
	B	0	1	9	4	2	4	6	3	1	6	2	2
	C	0	4	32	6	16	14	2	16	18	12	14	10
medium (in h)	A	0	17	29	13	16	17	23	21	3	20	19	7
	B	0	8	22	7	11	12	4	23	3	13	10	8
	C	0	27	41	22	23	23	24	32	11	25	24	19
large (in h)	A	1	37	46	25	29	30	21	33	30	32	26	25
	B	9	44	48	31	35	36	29	41	32	35	35	31
	C	14	48	48	36	37	38	35	42	33	38	36	36

In Figure 12, we see that the number of PSU candidates strongly influences the computation time. The problem is easy to solve when only one PSU candidate is considered. In this case, no decision has to be made about where to place PSUs, but only where to place ITUs. This makes the problem much easier. Increasing the number of PSU candidates increases the computation time significantly.

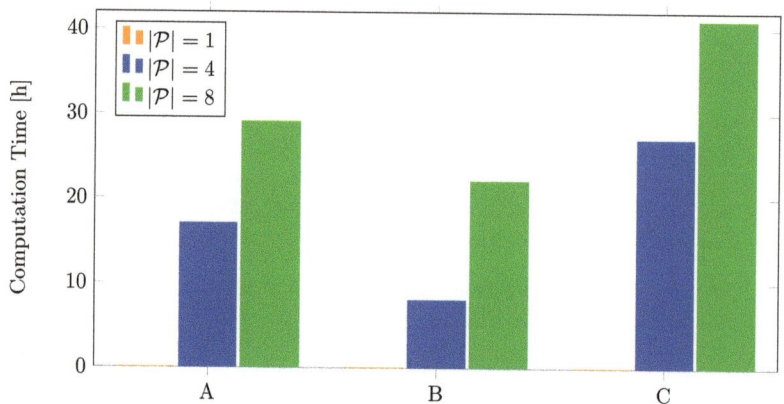

Figure 12. Influence of the number of PSU candidates on the computation time for medium sized instances of structures A, B, and C.

In Figure 13, we observe that increasing the number of service requests also leads to increased computation time. However, increasing the number of service requests leads to a much lower increase in the computation time than increasing the number of PSU candidates. On average, the increase from 30% to 70% is greater than the increase from 70% to 100%. The reason is the overlap of the requests. If all requests are included, an extremely high number of service requests is considered. Many of these requests have a high percentage of overlap. It is therefore conceivable that there are service requests that are already covered by the infrastructure requirements of other service requests. This means, for example, that if two service requests are fulfilled, a third request is automatically fulfilled as well. From the perspective of the problem, it makes no difference whether this third service request is considered or not. This effect will especially occur with a very large number of service requests. Thus, if we consider 70% of the requests, nearly all original requests are covered, and the difference between 100% is small. However, considering 30% of the requests will not cover all requests. Thus, the computation time increases substantially from 30% to 70% as more requests with no/few overlapping are added to the instance.

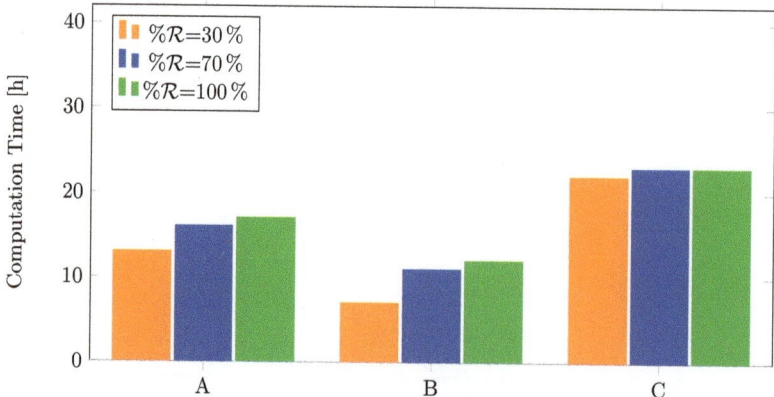

Figure 13. Influence of the percentage of considered requests on the computation time for large sized instances of structures A, B, and C.

Figure 14 shows the influence of the energy ratio on the computation time. For structure A of the medium instance, a ratio of 1.2 has the highest computation time, and for cases B and C, a ratio of 3.24. Table 8 also shows that the ratio of 3.24 mostly leads to the highest computation time. The combinatorics of the problem can explain this. An ei/ec ratio of 1.2 means that the energy intake is slightly larger than the energy consumption. Thus, almost the entire infrastructure has to be equipped with ITUs. There are few possibilities to realize this. The extreme case $ei = ec$ illustrates this: In order to supply sufficient energy to the vehicles, the complete infrastructure must be equipped with ITUs. The only decision to be made is which PSU is to be connected. Therefore, there are only $|P|$ different feasible solutions. If ei/ec is very large, then the energy intake is significantly greater than the energy consumption per distance unit of ITU passed by the vehicle. In this case, only a very small part of the infrastructure has to be equipped. For this reason, a very large part of the solutions can be excluded since they are too expensive. In the case of a medium-sized energy ratio, such as 3.24, many solutions are feasible and not too expansive, and many solutions can potentially be considered. For this reason, the computation time is usually the largest for this value.

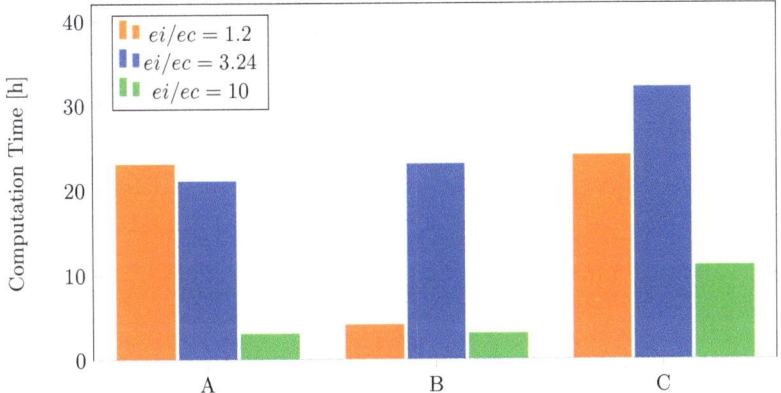

Figure 14. Influence of the energy ratio on the computation time for large sized instances of structures A, B, and C.

Figure 15 shows that increasing the c^{psu}/c^{itu} ratio leads to lower computation times. This means a high investment in PSUs compared to ITUs lead to low computation times. As already shown, lower computing times occur if fewer PSUs are considered. If the PSUs are very expensive compared to the ITUs, it is not economical to build many of them. Rather, more ITUs would be built instead of one additional PSU. The question of how many PSUs should be built is not as difficult if the PSUs are very expensive. However, if the PSUs are less expensive, it may be useful to build multiple PSUs. In this case, there is the question of where to place the PSUs and whether it is better to build additional ITUs instead of one additional PSU. These aspects increase the combinatorial difficulty of the problem and thus the computing times.

In a final study, we re-ran the instances with the features that led to the highest computation times, but now with a CPU time limit of 200 h (as opposed to the initial 48 h). Table 9 shows the results. Optimality could not be proven for any of the three instances, even after 200 h of computation. On the other hand, the remaining integrality gaps between 3.7% and 8.4% indicate that even for those hard instances, the Gurobi solver was able to find solutions that may be acceptable from a practical point of view as their worst-case deviation from optimality is not too large.

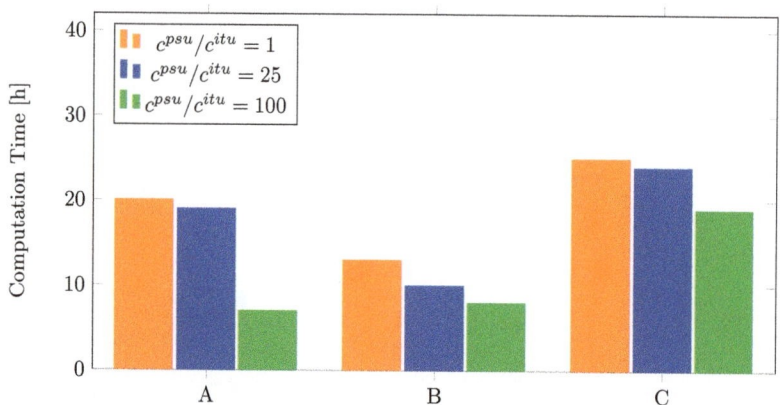

Figure 15. Influence of the investment ratio on the computation time for medium sized instances of structures A, B, and C.

Table 9. Computation of most difficult instances.

		$|P|$	%$|R|$	ei/ec	c^{psu}/c^{itu}	CPU [h]	Gap
large	A	8	100%	3.24	1	200	3.7%
	B	8	100%	3.24	1	200	8.4%
	C	8	100%	3.24	1	200	7.3%

Finally, there is the question of what these computational results mean for real problem instances. In real problem instances, there will probably be many possible PSU locations because it is conceivable that there is a connection to the grid at every gate position. Modeling all of them would most likely make the problem intractable. However, several PSU locations close to each other could easily be represented by one PSU candidate. The justification for this approach is that we observed a very large number of equally optimal solutions, so it is non-unjustified to hope that we can drop many PSU candidate positions without losing all the optimal solutions from the solution space.

We further assume that there will be a large volume of traffic on the apron in the future. This is why we are looking at a large number of service requests. Instead of operating with a two-trip structure as in our instance generation process, we could also create instances with single-trip structures. This would reduce the number of requests to consider, making the problem numerically more tractable. On the other hand, it would lead to more infrastructure solutions that would be more expensive but more robust by giving the vehicles more charging opportunities.

As mentioned in Section 4, the energy ratio of 3.24 is based on actual data from Broihan et al. [8]. They assumed a power of 100 kW and efficiency of 80% for the wireless power transfer. Since this is still a new technology and the application is very specific, real data for the investments are difficult to find. Most applications consider a single bus. Jeong et al. [28] mentioned 500 $/m as investment in a meter of an ITU and 50,000 $/unit for a PSU. However, the PSU needed for this application cannot be compared to the application at the airport, where many vehicles are powered simultaneously. Nevertheless, we can summarize that real-world problems mostly have the characteristics that make the problem particularly difficult. In addition, we have made simplifications even in the graphs of the large instance class since we do not consider all gates, depots and parking positions. Furthermore, we have chosen a still very large link length. It is conceivable that there are savings with smaller link lengths. With instances of real problem size, we can therefore assume even significantly higher computation times.

6. Conclusions

This paper presents an optimization model for planning an inductive charging infrastructure on airport aprons. The model is a simplified variant of the model presented by [8]. The paper aimed to investigate the problem properties that lead to high computation times when solving the model with standard solvers and the reasons behind these high computation times.

For this purpose, we systematically generated a large set of instances in three steps. At first, we created initial graphs based on real airport apron structures. Afterward, we adapted these initial graphs to the three defined instance classes small, medium and large. Finally, we created instances from each graph with different characteristics, such as the ratio of energy intake to energy consumption.

In the numerical evaluation, we showed that current commercial standard solvers such as Gurobi can find a feasible solution quickly. They can even find good and potentially very good solutions in an acceptable time. However, proving the optimality of a solution often takes very long. One reason for this is the high number of equally optimal solutions, which are, among other reasons, due to symmetries in the problem structure.

The computation times of instances resulting from graphs of similar size vary significantly and depend on the instance's attributes. Specific attributes lead to remarkably high computation times. We showed that an increasing number of service requests and PSU candidates lead to higher computation times, whereby the influence of the PSU candidates is more significant. In addition, we investigated the influence of the ratio of energy intake to energy consumption and the ratio of PSU investment to ITU investment. The energy ratio significantly determines the size of the charging infrastructure. The numerical investigations showed that, in particular, an energy ratio that leads to a medium-sized charging infrastructure also leads to high computing times, as in this case, there tend to be many equally good alternatives. The investment ratio of PSUs to ITUs influences whether an additional PSU or more ITUs must be built. Comparatively cheap PSUs lead to high computing times.

In real applications, the problem instances may be even more complex than those considered in this paper. We showed that for some instances, optimality could not be proven by standard solvers within a time limit of eight days. For this reason, it is crucial to investigate how the computation time can be reduced. One possibility is to break symmetries by extending the model with symmetry-breaking restrictions or considering undirected graphs. This study showed that the presented rather intuitive model formulation is very hard to solve. For this reason, one might search for alternative model formulations that lead to lower computation times. For example, a flow-based model formulation in which the PSUs serve as sources and the ITUs serve as sinks would be conceivable. Further, future research projects should consider the use of heuristics and customized solution approaches.

Further research could further investigate the interaction of the power dimensioning and the allocation of the charging infrastructure's components. The model presented here hence provides substantial opportunities for such extensions.

Author Contributions: Conceptualization, N.P., I.N., J.B. and S.H.; methodology, N.P.; software, N.P. and J.B.; validation, N.P., I.N., J.B. and S.H.; formal analysis, N.P., I.N. and J.B.; investigation, N.P., I.N. and J.B.; data curation, N.P., I.N. and J.B. ; writing—original draft preparation, N.P., I.N. and J.B.; writing—review and editing, S.H.; visualization, N.P., I.N. and J.B.; supervision, S.H.; project administration, S.H.; funding acquisition, S.H. All authors have read and agreed to the published version of the manuscript.

Funding: We would like to acknowledge the funding by the Deutsche Forschungsgemeinschaft (DFG, German Research Foundation) under Germany's Excellence Strategy—EXC 2163/1—Sustainable and Energy Efficient Aviation—Project-ID 390881007. The publication of this article was funded by the Open Access Fund of the Leibniz University Hannover.

Data Availability Statement: The data presented in this study are openly available in the Research Data Repository of the Leibniz University Hannover at https://doi.org/10.25835/itr694hg (accessed on 8 August 2022).

Acknowledgments: We acknowledge the support of the cluster system team at the Leibniz University Hannover, Germany, in the production of this work.

Conflicts of Interest: The authors declare no conflict of interest.

Abbreviations

The following abbreviations are used in this manuscript:

CPU	Central Processing Unit
DICP	Dynamic Inductive Charging Problem
DICP-MV	Dynamic Inductive Charging Problem considering Multiple Vehicle Types
ITU	Inductive Transmitter Unit
LP	Linear Program
PSU	Power Supply Unit
SOC	State of Charge

References

1. Bopst, J.; Herbener, R.; Hölzer-Schopohl, O.; Lindmaier, J.; Myck, T.; Weiß, J. *Umweltschonender Luftverkehr*; Für Mensch & Umwelt: Dessau-Roßlau, Germany, 2019.
2. Interreg CENTRAL EUROPE. *Electric Mobility Action Plan*; Interreg CENTRAL EUROPE: Vienna, Austria, 2019.
3. Flughafen München GmbH. Electric Vehicles at the Airport. Available online: https://www.munich-airport.com/electric-vehicles-at-the-airport-5938664 (accessed on 25 April 2022).
4. Royal Schiphol Group. Electric Transportation between Gate and Plane. Available online: https://www.schiphol.nl/en/schiphol-group/page/electric-transport-between-aircraft-and-gate/ (accessed on 25 April 2022).
5. Bulach, W.; Hacker, F.; Haller, M.; Minnich, L.; Weber, M.; Hofmann, M.; Siehler, E.; Salzer, S. *Der Weg zur vollelektrischen Flughafenflotte—2. Working Paper aus dem Projekt Scale Up!*; Öko-Institut e.V.: Berlin, Germany, 2020.
6. Panchal, C.; Stegen, S.; Lu, J. Review of static and dynamic wireless electric vehicle charging system. *Eng. Sci. Technol. Int. J.* **2018**, *21*, 922–937. doi: 10.1016/j.jestch.2018.06.015. [CrossRef]
7. Helber, S.; Broihan, J.; Jang, Y.; Hecker, P.; Feuerle, T. Location Planning for Dynamic Wireless Charging Systems for Electric Airport Passenger Buses. *Energies* **2018**, *11*, 258. [CrossRef]
8. Broihan, J.; Nozinski, I.; Pöch, N.; Helber, S. Designing Dynamic Inductive Charging Infrastructures for Airport Aprons with Multiple Vehicle Types. *Energies* **2022**, *15*, 4085. [CrossRef]
9. SAE International. *Wireless Power Transfer for Light-Duty Plug-in/Electric Vehicles and Alignment Methodology*; SAE International: Warrendale, PA, USA, 26 August 2022.
10. Li, S.; Mi, C.C. Wireless Power Transfer for Electric Vehicle Applications. *IEEE J. Emerg. Sel. Top. Power Electron.* **2015**, *3*, 4–17. [CrossRef]
11. Cirimele, V.; Diana, M.; Freschi, F.; Mitolo, M. Inductive Power Transfer for Automotive Applications: State-of-the-Art and Future Trends. *IEEE Trans. Ind. Appl.* **2018**, *54*, 4069–4079. [CrossRef]
12. Lukic, S.; Pantic, Z. Cutting the Cord: Static and Dynamic Inductive Wireless Charging of Electric Vehicles. *IEEE Electrif. Mag.* **2013**, *1*, 57–64. [CrossRef]
13. Ahmad, A.; Alam, M.S.; Chabaan, R. A Comprehensive Review of Wireless Charging Technologies for Electric Vehicles. *IEEE Trans. Transp. Electrif.* **2018**, *4*, 38–63. [CrossRef]
14. Covic, G.A.; Boys, J.T. Modern Trends in Inductive Power Transfer for Transportation Applications. *IEEE J. Emerg. Sel. Top. Power Electron.* **2013**, *1*, 28–41. [CrossRef]
15. Imura, T.; Hori, Y. Maximizing Air Gap and Efficiency of Magnetic Resonant Coupling for Wireless Power Transfer Using Equivalent Circuit and Neumann Formula. *IEEE Trans. Ind. Electron.* **2011**, *58*, 4746–4752. [CrossRef]
16. Moon, S.; Kim, B.C.; Cho, S.Y.; Ahn, C.H.; Moon, G.W. Analysis and Design of a Wireless Power Transfer System With an Intermediate Coil for High Efficiency. *IEEE Trans. Ind. Electron.* **2014**, *61*, 5861–5870. [CrossRef]
17. Ko, Y.D.; Jang, Y.J. The Optimal System Design of the Online Electric Vehicle Utilizing Wireless Power Transmission Technology. *IEEE Trans. Intell. Transp. Syst.* **2013**, *14*, 1255–1265. [CrossRef]
18. Hwang, I.; Jang, Y.J.; Ko, Y.D.; Lee, M.S. System Optimization for Dynamic Wireless Charging Electric Vehicles Operating in a Multiple-Route Environment. *IEEE Trans. Intell. Transp. Syst.* **2018**, *19*, 1709–1726. [CrossRef]
19. Ko, Y.K.; Oh, Y.; Ryu, D.Y.; Ko, Y.D. Optimal Deployment of Wireless Charging Infrastructure for Electric Tram with Dual Operation Policy. *Vehicles* **2022**, *4*, 681–696. [CrossRef]
20. Mensen, H. *Handbuch der Luftfahrt*; Springer: Berlin/Heidelberg, Germany, 2013. [CrossRef]

21. Broihan, J. Models and Algorithms for Designing Dynamic Inductive Charging Infrastructures on Airport Aprons. Unpublished Dissertation, 2022.
22. Jang, Y.J. Survey of the operation and system study on wireless charging electric vehicle systems. *Transp. Res. Part Emerg. Technol.* **2018**, *95*, 844–866. [CrossRef]
23. Majhi, R.C.; Ranjitkar, P.; Sheng, M.; Covic, G.A.; Wilson, D.J. A systematic review of charging infrastructure location problem for electric vehicles. *Transp. Rev.* **2021**, *41*, 432–455. [CrossRef]
24. Yatnalkar, G.; Narman, H. Survey on Wireless Charging and Placement of Stations for Electric Vehicles. In Proceedings of the 2018 IEEE International Symposium on Signal Processing and Information Technology (ISSPIT), Louisville, KY, USA, 6–8 December 2018; pp. 526–531. [CrossRef]
25. Jang, Y.J.; Jeong, S.; Ko, Y.D. System optimization of the On-Line Electric Vehicle operating in a closed environment. *Comput. Ind. Eng.* **2015**, *80*, 222–235. [CrossRef]
26. Liu, Z.; Song, Z. Robust planning of dynamic wireless charging infrastructure for battery electric buses. *Transp. Res. Part Emerg. Technol.* **2017**, *83*, 77–103. [CrossRef]
27. Schwerdfeger, S.; Bock, S.; Boysen, N.; Briskorn, D. Optimizing the electrification of roads with charge-while-drive technology. *Eur. J. Oper. Res.* **2022**, *299*, 1111–1127. [CrossRef]
28. Jeong, S.; Jang, Y.J.; Kum, D. Economic Analysis of the Dynamic Charging Electric Vehicle. *IEEE Trans. Power Electron.* **2015**, *30*, 6368–6377. [CrossRef]

Article

Design and Implementation of a Wireless Charging System Connected to the AC Grid for an E-Bike

Emin Yildiriz [1] and Murat Bayraktar [2,*]

1. Department of Electrical and Electronics Engineering, Faculty of Engineering, Düzce University, Düzce 81620, Turkey; eminyildiriz@duzce.edu.tr
2. R&D Department, Farba Automotive, Kocaeli 41420, Turkey
* Correspondence: murat.bayraktar@farba.com.tr; Tel.: +90-262-781-81-18

Abstract: This paper aims to design an IPT for wireless charging of an e-bike and to control the charge of the e-bike from the primary-side. Optimum IPT design has been made according to the 36 V battery bank requirements. The no-load condition test has been performed before charging started in the IPT system connected to the AC grid. The primary-side DC-link voltage of 4–5 V required for this test is provided by the designed forward converter. The charge control has been also made from the forward converter on the primary-side. For this, the forward converter's operation in peak current mode (PCM) has been used. Finally, a prototype has been implemented that works at a maximum DC/DC efficiency of 87.52% in full alignment and 83.63% in 3 cm misalignment. The proposed control algorithm has been tested in this prototype at different load stages.

Keywords: inductive power transfer; wireless charging; e-bikes; forward converter

1. Introduction

The research about plug-in and contactless charging of e-vehicles is increasing with the increase in the use of e-vehicles all over the world [1]. Depending on the power level and the usage area of e-vehicles, static or dynamic wireless charging can be used. In these power transmission systems, which are loosely coupled due to the height of the vehicle sub-chassis from the ground, the compensation is used on the primary side and secondary side to increase the power transmission efficiency. The main compensation topologies such as Series-Series, Series-Parallel (SP), and also the popular topologies that have additional components such as Inductor/Capacitor/Capacitor (LCC) [2], Inductor/Capacitor/Inductor (LCL) [3,4] etc. have been applied on the primary and secondary sides. The SS topology was preferred in this paper, since it is frequently chosen in the literature at similar power levels and uses fewer components. The other important components in the IPT structure are converter structures. In an IPT system supplied from a 50–60 Hz AC grid, the dual-stage converters are traditionally preferred on the primary side (Figure 1). Particularly in recent years, there have been researchers who prefer the single-stage converter on the primary side [5,6]. By using the matrix converter on the IPT, a more compact primary side is provided independently of the lifetime and cost of the DC link capacitor. However, the dual-stage converter can be designed and controlled independently at each stage and also operates with high efficiency even at variable loads. Thus, this topology has been preferred in many industrial applications [7].

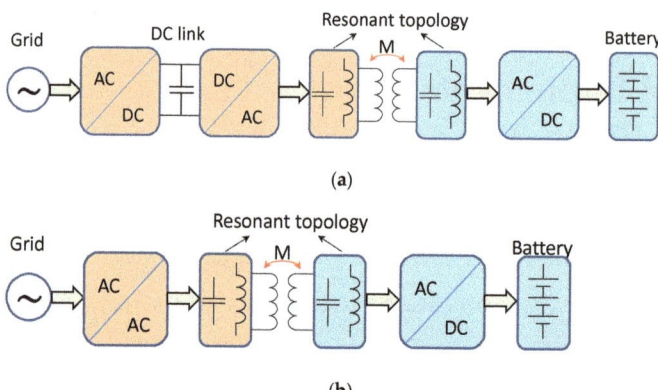

Figure 1. Primary side configurations of IPT: (**a**) Dual-stage; and (**b**) Single-stage.

The battery characteristics should be taken into account when designing an IPT for the electrical device to be wirelessly charged. For the battery to operate in a healthy and long-lasting manner, it must be charged in Constant Current (CC) and Constant Voltage (CV) mode according to the battery charge profile [8]. During the charging process, the equivalent load resistance (R_o) is not constant as the charging current and voltage change. Although the mutual inductance between the coils (M) is constant in the full-alignment state, the transferred power, efficiency, charging current, and voltage also change since R_o changes.

The charging current and voltage can be controlled from the primary side, the secondary side, or both sides in case of misalignment or load changes. If the control is undertaken on the secondary side, the desired control can be achieved with the current and voltage data of the load. The control is achieved with an added DC/DC converter or using a controlled rectifier on the secondary side [9,10]. That is, the control from the secondary side does not require communication components but creates a more complex secondary side. When performing CC/CV control from the primary side, the variation of the load current and voltage must be known or estimated. These data can be transferred from the secondary side to the primary side via a wireless communication link [11,12]. However, the data transmission security may be compromised in cases such as misalignment or the presence of any foreign objects. The parameter estimation approach can be based on the load estimation [13,14] and/or the mutual inductance estimation [15]. The CC/CV control from the primary side is generally based on operating the H-bridge inverter with variable frequency or the phase-shift method [16,17]. The variable frequency operation may affect the power transfer performance and the efficiency due to divergence from the resonance frequency. The phase-shifting method is the most widely used [4,18]. However, the phase-shifting method does not allow the use of a standard soft-switched inverter. Moreover, achieving the secondary side control with the phase-shifting method on the primary side as in the e-bike application focused on in this study (an IPT with a DC-link voltage of 400 V on the primary side and also a low voltage on the secondary side) is very difficult. This is because the change sensitivity of the phase-shift angle must be very high. Therefore, the sensitivity of the phase-shifting limits the secondary side DC link voltage. Consequently, a DC/DC converter is required for the CC/CV control from the primary side of IPT applications, such as the wireless charging developed for the e-bike and sourced from a 230 V AC grid.

Nowadays, the research on wireless charging of e-vehicles is generally focused on a power level of 3.3 kW and above. The primary side and secondary side DC-link voltages are usually 400 V in IPT designs for these power levels [19–22]. Whereas the primary side DC-link voltage is determined according to the AC grid, the secondary side DC voltage is related to the battery bank voltage. The power level to be transferred in the charging of e-bikes and the voltage level of the battery group are low. In e-bike applications encountered in the literature, the system responses have been examined by connecting a DC supply to the inverter input [23–29], or a DC-link voltage in the range of 60–85 V can be achieved on the primary side using a grid-connected step-down transformer [30]. In these papers, the CC/CV control was carried out with the phase-shifting method over the h-bridge inverter. Using a DC/DC converter instead of a step-down transformer at the inverter input provides a more compact primary pad. A dual-stage converter was used in [31], and a buck converter was preferred at the inverter input. Due to the operation limits of the buck converter, this wireless charging system requires the use of a DC/DC converter on the secondary side for the charge control. In this case, both the use of more components has emerged and the idea of a simple secondary side acquisition has been moved away from. Considering the researches on wireless charging of e-bikes that use low secondary side DC-link voltage, the main contributions of this paper are summarized as follows.

(1) A wireless charging system connected to the AC grid has been designed to meet the battery charging requirements of the e-bike.
(2) The IPT and AC grid are isolated from each other by the use of a forward converter. At the same time, the low primary side DC-link voltage required testing the presence of load and large misalignment can be provided with the forward converter.
(3) The CC/CV control is achieved with a forward converter working in PCM mode instead of an H-bridge inverter. Thus, the soft-switching methods can be easily applied in an inverter operating at a constant frequency and duty-ratio.

The organization of this paper is as follows. The optimum IPT design was carried out considering the charging requirements of the e-bike, as described in Section 2. Then, the design of the forward converter, in which the charge control and the no-load condition test were carried out, is presented. The simulation and experimental results are presented in Section 4, comparatively. Finally, Section 5 concludes this paper.

2. Design of Inductive Power Transfer System for E-Bikes

The limits of electrical parameters such as DC-link voltages and resonance frequency, as well as physical criteria such as the air-gap of the power transfer, should be determined correctly when starting an IPT system design. Wireless charging of a 250 W e-bike with 36 V, 20 Ah gel batteries was investigated in this paper. The charging system's input is connected to a 230 V AC grid. The general scheme of the system is given in Figure 2. According to the battery characteristics, the maximum voltage needed to charge the 3 × 12 V battery pack is 44 V. When charging at a constant current, the charging current is required to be 2.5 A.

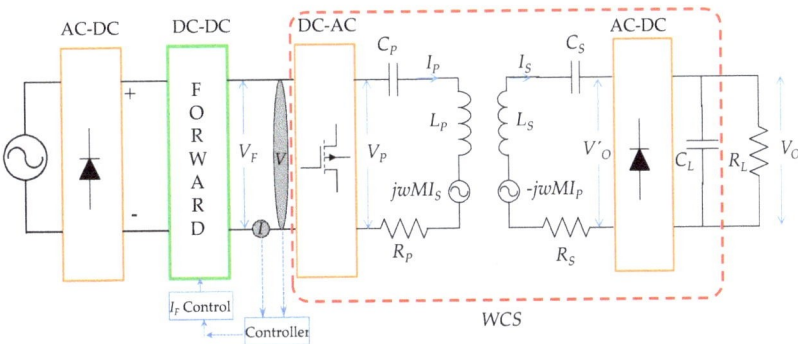

Figure 2. Structural diagram of the IPT proposed for e-bike wireless charging.

The secondary side output voltage of the IPT, V_o' is calculated using Equation (1) according to the battery charge requirements. The primary and secondary side voltages are expected to be close to each other for optimum coil usage [32]. The primary side voltage V_P was selected considering this closeness relation. There is a relationship between the output voltage of the DC-DC converter (V_F) and V_P as in Equation (2). Accordingly, a forward converter was preferred in this study in order to reduce the output voltage of the rectifier connected to the AC grid to the desired DC voltage level.

$$V'_o = \frac{\pi}{2\sqrt{2}} V_o \tag{1}$$

$$V_P = \frac{4}{\pi\sqrt{2}} V_F \tag{2}$$

The voltage equations of the primary and secondary side are written as Equations (3) and (4) for SS topology. Here ω is the resonance frequency. The self-inductances (L_P and L_S) of the primary and secondary side windings depend on the winding dimensions and the number of turns, which are decided according to the coil design. Primary and secondary resonance capacitors (C_P and C_S) are determined considering L_P, L_S, and ω. Another important parameter in voltage equations, R_o, is the equivalent resistance at the rectifier input and is calculated from Equation (5). The primary and secondary sides are expected to operate in resonance during the power transmission. Thus, the efficiency of the power transfer can be calculated from (6). Using Equation (4), the ratio between primary and secondary side currents is written as in (7). Equation (8) is obtained when this ratio is used in Equation (6). Consequently, the change in load and mutual inductance directly affects the power transfer efficiency (PTE).

$$V_P = R_P I_P + j\left(L_P \omega - \frac{1}{C_P \omega}\right) I_P + j\omega M I_S \tag{3}$$

$$j\omega M I_P = R_S I_S + j\left(L_S \omega - \frac{1}{C_S \omega}\right) I_S + R_o I_S \tag{4}$$

$$R_o = \frac{8}{\pi^2} R_L \tag{5}$$

$$PTE = \frac{R_o I_S^2}{R_P I_P^2 + R_S I_S^2 + R_o I_S^2} \tag{6}$$

$$\frac{I_P}{I_S} = \frac{R_S + R_o}{\omega M} \tag{7}$$

$$PTE = \frac{R_o}{R_P \left(\frac{I_P}{I_S}\right)^2 + (R_S + R_o)} = \frac{R_o}{R_P \left(\frac{R_S + R_o}{\omega M}\right)^2 + (R_S + R_o)} \tag{8}$$

The resonance frequency and mutual inductance should be high for maximum PTE, as seen in Equation (9). However, as the operating frequency increases, the effective series resistance (ESR) of the coils increases too. Therefore, quality factors should also be taken into account in the PTE calculation. The quality factors of the primary and secondary sides must be carefully selected to avoid bifurcation during charging. The quality factors are calculated as in Equation (10) for the SS topology. Q_P and Q_S values are usually chosen to be close to each other and to be $Q_P > Q_S$ [33,34].

$$\omega M \gg \sqrt{R_P(R_S + R_O)} \tag{9}$$

$$Q_P = \frac{L_P R_O}{\omega M^2}, \quad Q_S = \frac{\omega L_S}{R_O} \tag{10}$$

In the optimum IPT design, besides electrical constraints such as input and output voltages and required power, the physical constraints should also be taken into account. The maximum outer dimensions of the coils and power transmission height are known, due to the sub-chassis size limitation. In order to protect the secondary winding fixed to the sub-chassis of the e-bike, a plastic protective cover according to the winding dimensions was prepared. The height of the impact-proof plastic cover was 10 cm from the ground. The maximum area that the secondary winding could use was 240 × 280 mm, considering the sub-chassis limits. In order to make the most of this area, a rectangular winding pair was preferred. In this study, a coil pair with an air core was chosen in order to avoid additional weight on the bicycle. Thus, since the winding inductances can be calculated analytically, not FEA, the optimum winding pair could be determined quickly.

In addition to physical constraints, design constraints such as winding current densities, maximum frequency, and the avoidance of bifurcation should also be considered. It is possible to use smaller winding pairs for the same power transfer as the operating frequency increases. However, the magnetic flux density outside the power transmission region also increases with the operating frequency. The maximum operating frequency of the optimal IPT to be designed was selected at around 85 kHz, considering compliance with ICNIRP and IEEE c95.1 standards and SAE J2954 criteria. The maximum winding current density was determined as 3 A/mm^2 in order to keep the thermal effect small.

The winding designs capable of transmitting the desired power were scanned with an algorithm as in [33], considering the determined design constraints and physical constraints. The self and mutual inductances of the air-core windings were calculated analytically. The winding inductances depend on the winding dimensions and the number of turns. The K_D parameter was used to determine the best winding pair among the winding pairs that could transfer the desired power. The K_D parameter is a design factor, which is determined by the winding quality factors and the maximum operating frequency, and defined as the winding utilization factor [35].

The parameters of the optimum IPT are given in Table 1. The windings were wound with 38 AWG litz wire, taking into account the operating frequency and the current density. When a load close to the equivalent load was connected to the secondary side, the current and voltages close to the design values were measured experimentally. In the experimental study, the winding resistances were high due to the additional connection lengths. This situation was also reflected in the observed experimental efficiency.

Table 1. Design parameters of the optimum IPT.

Parameter	MATLAB	Measured
R_P (Ω)	0.3698	0.433
R_S (Ω)	0.3501	0.4067
L_P (µH)	205.1179	203.1
L_S (µH)	189.9593	187.31
M (µH)	26.4054	25.5
C_P (nF)	17.1281	17.447
C_S (nF)	18.4895	18.791
V_P (Volt)	41.1	41.1
I_P (A)	2.9087	2.94
I_S (A)	2.8465	2.85
f_0 (kHz)	84.9234	85.19
P_o' (Watt)	113.5933	113.145
η (%)	95.012	93.18

The load in the designed IPT system is not static. The equivalent resistance value will also change with the battery charge level. The current and voltage gains for different loads were calculated according to the design parameters. Due to the nature of the SS topology, an uncontrolled IPT system operates in CC mode. Since the magnitudes of the current-voltage gain G_{I-V} are small, a limited change in current is observed in the load changes that may occur while operating at the resonance frequency (as seen in Figure 3a). When the voltage gain for load changes is examined, it is seen that G_{V-V} increases significantly as the resistance value increases at 85 kHz, for which the IPT is designed (Figure 3b). In other words, due to the nature of the SS topology, it cannot give a constant voltage output at variable loads. Therefore, especially when the secondary side is open-circuit, the output voltage can reach dangerous points.

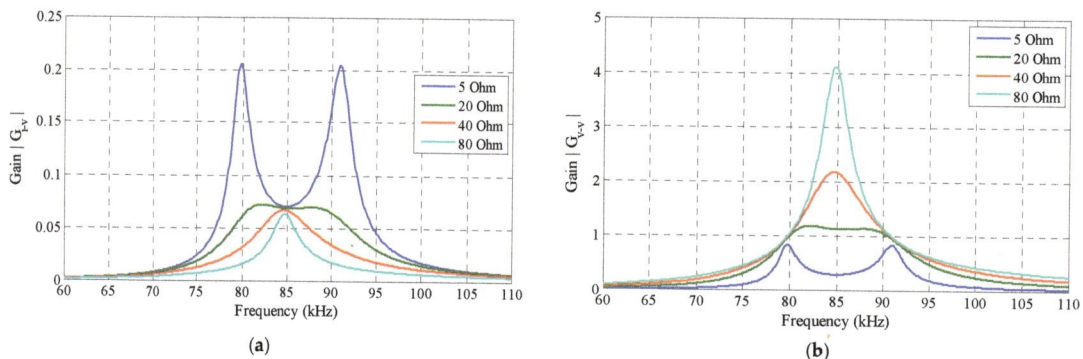

Figure 3. Frequency responses of the designed IPT system: (a) G_{I-V}; (b) G_{V-V}.

3. Forward Converter Design

The forward converter isolates the input and output sides of the converter from each other using the forward transformer. 300 V DC voltages can be stepped down to 5–10 V using a forward converter. Considering the switching times of the semiconductor circuit elements, it is difficult to obtain such a duty cycle with a buck converter. Although the power levels of forward converters are limited, a forward converter can be used in cases in which the required power is low (44 V × 2.5 A), as in this study.

In the two-switch forward converter, the current and voltage stress that the switches are exposed to is less than that of a single-switch one. The equivalent circuit of the two-switch forward converter according to the ON and OFF operating modes is shown in Figure 4. When the switches (Q1 and Q2) are ON, energy is transferred depending on the conversion ratio from primary to secondary of the transformer. At this moment, D3 is ON, whereas D4 is OFF. When the switches turn off, the primary winding of the transformer is reversely connected to the input voltage via D1 and D2. Thus, there is no need to use an additional winding to reset the transformer. On the secondary side, the voltage of L_F reverses. Therefore, D3 turns off and D4 turns on. Thus, the linearly decreasing current of L_F continues to flow. The output voltage of the forward converter is calculated from Equation (11). Here η_F, N_F, D denote the converter efficiency, forward-turn ratio and duty ratio, respectively. The turn ratio of the transformer is calculated taking into account the maximum of the duty ratio and the minimum input voltage. L_F and C_F are calculated according to the limited current and voltage ripple.

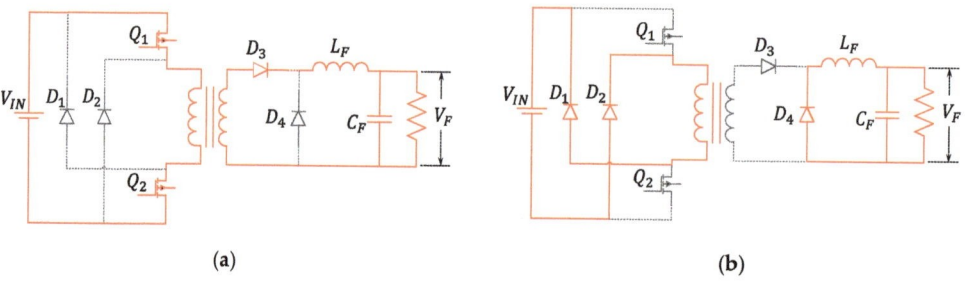

Figure 4. The equivalent circuits of the two-switch forward converter: (a) ON state; (b) OFF state.

The voltage ripple of the converter output will further increase the voltage ripple in the charging process. Therefore, the magnitude of the capacitor at the converter output is important (12). Here ω_F is the forward switching frequency, and ΔI_F and ΔV_F represent the ripple in current and voltage of the forward converter output, respectively. The input of the forward converter is connected to a single-phase full-wave rectifier. The parameters of the designed forward converter are given in Table 2.

$$V_F = \eta D V_{IN} N_F \qquad (11)$$

$$C_F = \frac{\Delta I_F}{\omega_F \Delta V_F} \qquad (12)$$

Table 2. The parameters of the forward converter.

ΔI_F	250 mA	D	% 15–47
ΔV_F	10 mV	N_F	1.63
V_{IN}	310 V	η	% 96
C_F	1000 uF	L_F	330 uH

4. The Control Strategy of No-Load Condition and Charging

4.1. No-Load Condition Control

Two separate controls work at the same time in the developed IPT. These are no-load condition control and charge control. Before starting charging, the presence of the load is tested. For this, the duty cycle of switches Q1 and Q2 is set to D_{test} so that the output voltage of the two-switch forward converter is equal to V_{F_test}. At this moment, the current of the forward converter I_F, is measured. If I_F is not less than I_{F_max1}, the system is kept in the load-test section. I_{F_max1} is the limit current value for the primary side's safe operation. If $I_F = I_{F\_max1}$, this could be for two reasons. First, it could be a secondary side open circuit. In the other case, there may be a fully charged battery on the secondary side. In both cases, it is kept as $V_F = V_{F\_test}$. The no-load condition control also works during the charging process. Thus, if the load is disconnected during charging, the primary side is protected. The no-load condition control is shown as Section-I in the control algorithm (Figure 5).

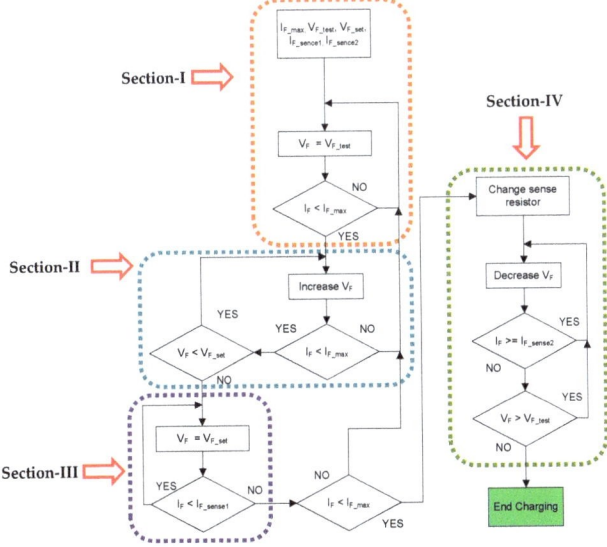

Figure 5. Flow chart of the control algorithm for wireless charging.

4.2. CC-CV Control Strategy

D is gradually increased starting from the D_{test} value, after the no-load condition test is passed. Thus, V_F also increases. In each step, it is checked whether the forward current exceeds the allowable upper limit. The increase in D continues until $V_F = V_{F\_set}$. Thus, $V_{F\_set} = 60$ V value to be applied for charging in constant current mode is reached. This part is shown as Section-II in the control algorithm.

When $V_F = V_{F\_set}$, and I_F has not reached I_{F_sense} yet, the equivalent impedance value of the battery is quite low and charging has started in CC mode. The nature of SS topology tends to keep the secondary side current constant. The secondary side output voltage increases with the increase in state of charge (SOC). At this moment, if it is on the primary side, I_F will increase. The charging process continues in CC mode until the $I_F = I_{F\_sense1}$ condition is met. When $I_F = I_{F\_sense1}$, the charging process has reached the stage of switching to CV mode in CC mode. This part of the control strategy is shown in Figure 5 as Section-III.

In the CV mode of the charging process, the equivalent resistance of the battery bank tends to rise rapidly. The limit current value must be reduced to the second peak current level (I_{F_sense2}) so that I_F current does not quickly reach the maximum value. As the charging process continues, I_F tends to increase. After this point, V_F is reduced to prevent the increase in I_F. For this, D is gradually reduced. At each step, it is checked whether I_F reaches the second peak current value. This loop continues until $V_F = V_{F\_set}$ (Section-IV). If this condition is met, the battery is now full and the charging process is completed.

5. Simulation and Experimental Results

The IPT parameters obtained in Table 1 were used to simulate the whole system on LTspice. The magnetic coupling (k) between L_P and L_S, and the equivalent load, are defined as variables in the simulation model of the whole system. k is determined from Equation (13) for the fully aligned coil pair. In order to accurately model and observe the system response, the open-loop response of the wireless charging system was first examined. Then, the developed controller performance was observed. The load parameter in the simulation was changed at 100 ms, 180 ms, and 240 ms of the simulation, considering the load stages used in the experimental study.

$$k = \frac{M}{\sqrt{L_P L_S}} \tag{13}$$

The experimental studies were carried out on a multi-stage and programmable load group to observe the response of the controller algorithm since the charging time of the batteries took a long time. Figure 6 shows this load group and the designed IPT system in full alignment. Experimental measurements were made at the 100 mm distance using the fully aligned coil pair.

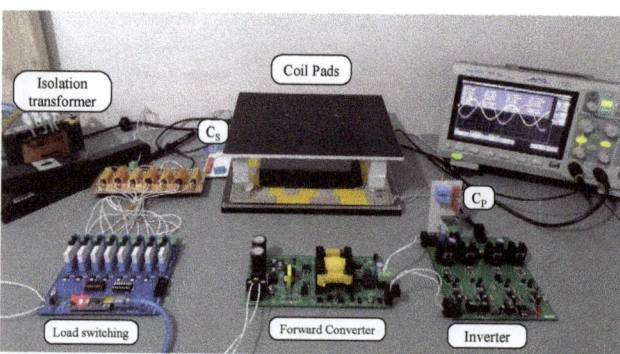

Figure 6. Experimental setup of the primary-side controlled IPT.

5.1. Open Loop System Response

While switching from CC mode to CV mode, V_o voltage is 44 V and I_o current is 2.5 A. At this point, the R_L equivalent battery resistance is 17.6 Ω. The IPT system must operate in CC mode below the equivalent load, and in CV mode above it. The R_L magnitudes to be used in the simulation were selected by considering the nominal values of the resistors used in the experimental study. Accordingly, R_L was set to 8.5 Ω, 10.5 Ω, and 14 Ω, respectively, to examine the CC mode. In this mode, V_F was kept constant at 60 V.

According to the experimental results, as the equivalent resistance increases, the load voltage V_o also increases (Figure 7a). The load current I_o is not affected by the load change as a general response of the SS topology (Figure 7b). V_o and I_o seen in the simulation results also coincide with the experimental results (Figure 7c,d).

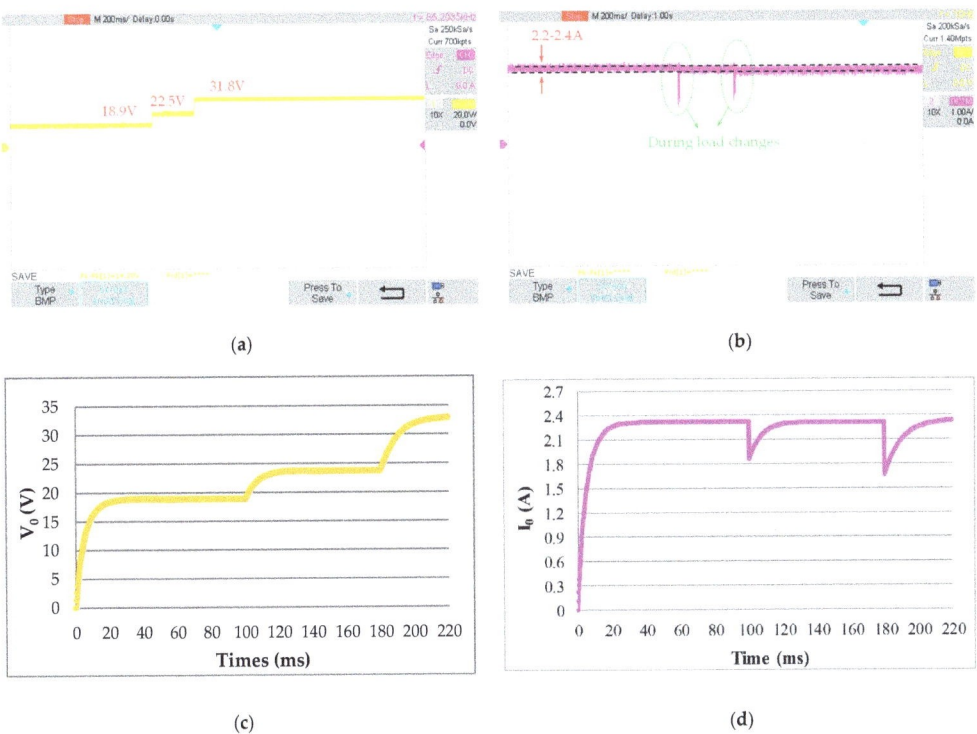

Figure 7. (a) measured V_O; (b) measured I_O; (c) simulated V_O; (d) simulated I_O when R_L is 8.5 Ω, 10.5 Ω, and 14 Ω, respectively.

The range of the R_L change is 17.6–440 Ω in CV mode. As SOC increases, the equivalent R_L will also increase too. In this case, the load power increases with the increasing output voltage. Therefore, the primary side current I_P will increase. The forward voltage was set to 40 V in this section of the experiment in order to protect the circuit elements. The load was increased from 16 Ω to 22 Ω, 28 Ω, and 32 Ω for the test in the CV section. As seen in Figure 8a, V_o continues to increase. For R_Ls in the CV section, the simulation model gives close responses to the experimental results (Figure 8b). If the variation of I_P is examined depending on R_L, it is seen that the primary side current magnitude doubles its nominal value at 32 Ω, although a V_F is applied below the nominal operating voltage (Figure 8c). Considering that the equivalent resistance value will approach 440 Ω when the battery is fully charged, it is obvious that the primary side current must be controlled.

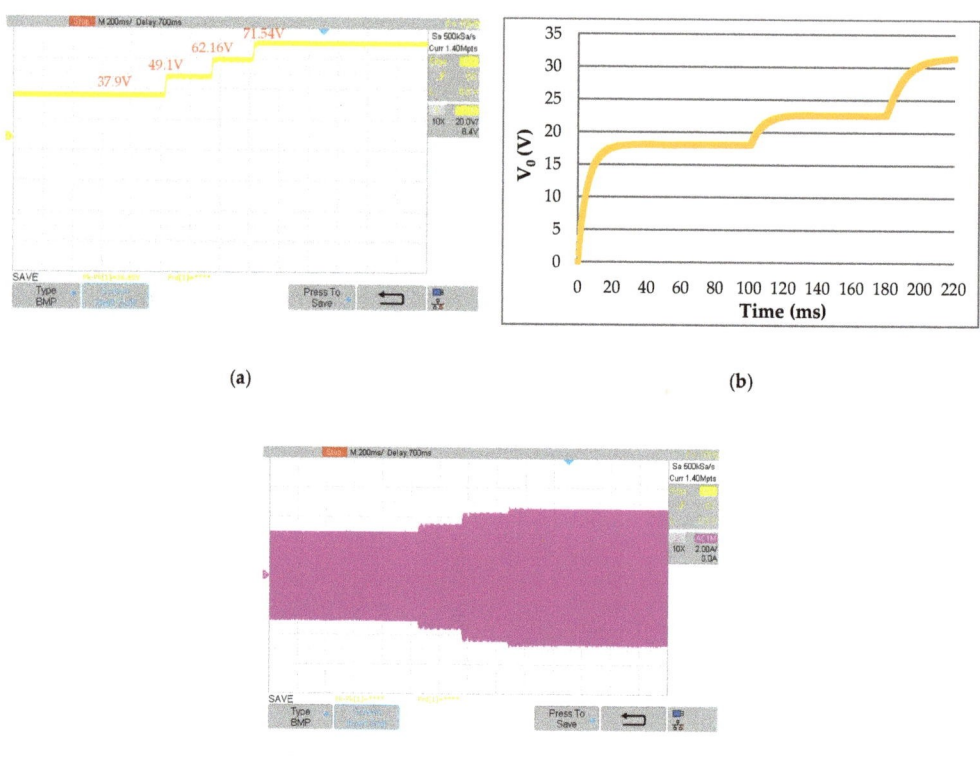

Figure 8. (**a**) measured V_O; (**b**) simulated V_O; (**c**) measured I_P when R_L is 16 Ω, 22 Ω, 28 Ω, and 32 Ω, respectively.

5.2. Closed Loop System Response

Switched load magnitudes were selected as 12.8 Ω, 17 Ω, 22.6 Ω, and 32 Ω to observe the overall system response in CC and CV modes. The output voltage of the forward converter was fixed at 60 V for the system that passed the test section of the algorithm (Figure 9a). It was shown before in Figure 7b that I_o did not change due to the SS topology effect in resistance changes up to 14 Ω. Therefore, a single resistance level of 12.8 Ω was considered sufficient to show the charge in CC mode in this section. The charging took place in CC mode in the first part of the control algorithm.

When the resistance increases from 12.8 Ω to 17 Ω, the output voltage V_O increases (Figure 9b). In this case, since the forward current will tend to increase rapidly, the voltage V_F is decreased by the control algorithm. However, a slight decrease was observed in I_O. While the resistance is 22.6 Ω, the charge continues in the CV section. I_O decreases with the increase in SOC (Figure 9c). The output voltage was kept constant by reducing V_F. When the resistance was increased from 22.6 Ω to 32 Ω, V_O was kept constant, and the I_O charging current continued to be decreased. If the resistance magnitude continued to increase at this stage of the charge, it would continue to be throttled until V_F was equal to V_{F_test}.

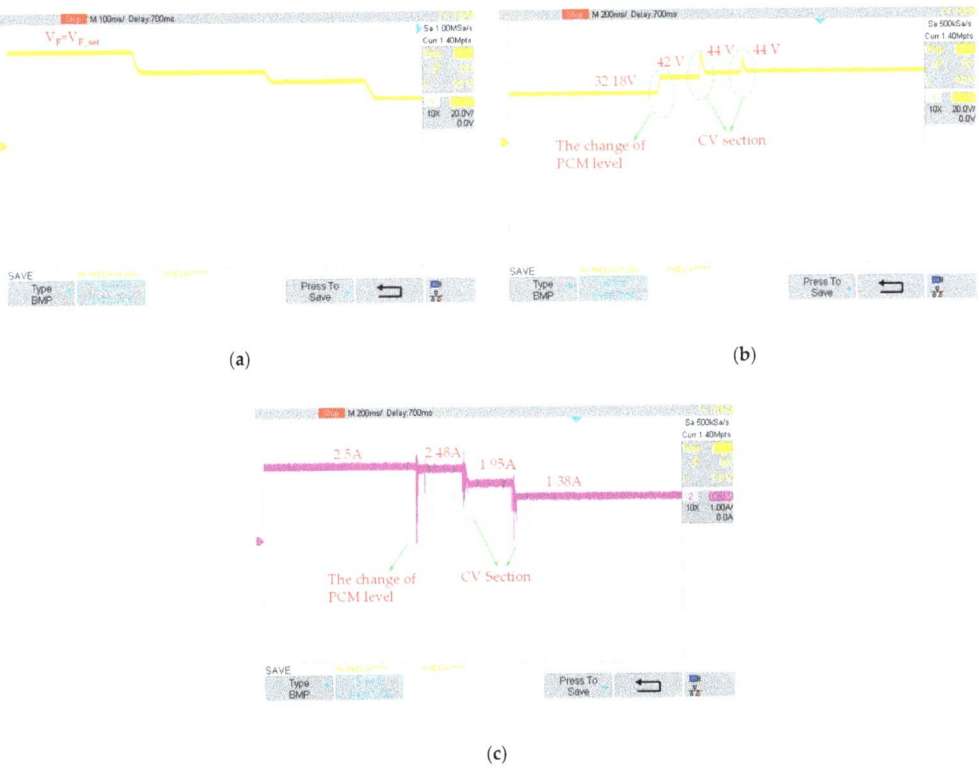

Figure 9. Experimental results of the control algorithm: the change of (**a**) V_F; (**b**) V_O; (**c**) I_O.

In the experimental results, there is an overshoot in the V_o waveform and an oscillation in the I_o waveform due to the sudden change in the load level. In the actual battery charge, these fluctuations will not be in question since there is not be such a sudden change in the equivalent resistance of the battery according to the SOC. In fact, these sudden changes can in practice be caused by a sudden change in the mutual inductance due to the occurrence of misalignment during charging or an object entering between the coil pads. These situations can be identified by detecting high-amplitude inrush current changes.

The primary side should be protected by reducing the forward converter output voltage, if the secondary side is open-circuit at any stage of the charge. The load resistance was increased from 30 Ohm to 440 Ohm for observing this situation in the LTspice model. The variation of V_F with the change of the resistance in the forward converter is shown in Figure 10. The V_F voltage is reduced to 4.041 V. At this moment, I_F also reaches the maximum allowable value. The variation of the current passing through the inductance L_F in the forward converter with the switching of the load is shown in Figure 10b.

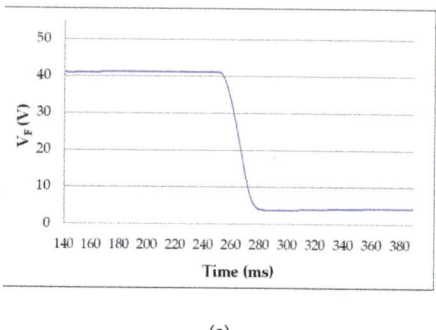 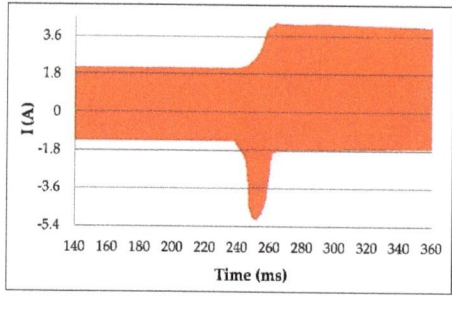

Figure 10. (a) V_F; (b) the current of L_F when disconnecting from the load.

The wireless charging system for the e-bike developed in this paper was compared with the same power level IPTs in terms of efficiency. As can be seen in Table 3, the IPT proposed in this study provides the desired power transmission with higher efficiency at a higher transmission distance than its counterparts. Moreover, since the DC-link voltage on the primary side is high, it can be directly connected to the AC grid.

Table 3. The comparison of WCSs of similar power level.

	Topology	Primary Side Voltage (V)	Secondary Side Voltage (V)	Power (W)	Distance (mm)	Freqency	Eff. (%)
[3]	LCL-LCL	50	50	100	180	1 Mhz	73.6
[17]	S-S	48	18	53,79	30	85.5 kHz	81
[18]	LCL-S	48	26	78,31	35	85 kHz	84.103
[23]	S-S	20	20	90	-	95.6 kHz	90.3
[25]	S-S	-	24	100	30	-	79
[26]	S-S	24	48	250	200	200 kHz	92.15
[27]	S-S	72	48	100	30	100 kHz	-
[29]	S-S	48	36	-	30	100 kHz	-
[29]	S-LCC	48	20	55	-	200 kHz	-
[30]	S-S	~21 V	44,5	84	-	85 kHz	93.17
[31]	S-S	400 V	42 V	100	70	83.1 kHz	80
Prop.	S-S	310 V	44 V	110	100	85 kHz	87.52

6. Conclusions

A wireless charging system, which uses a dual-stage configuration and is supplied from a 230 V AC grid, was designed for charging the 36 V battery bank of an e-bike in this study. An optimum IPT for wireless charging was developed using the dimensions and components obtained in the analytical design. Developed in the SS resonance topology, the current on the primary side of the IPT can reach dangerous values if the charge is in CV mode, the secondary side is open circuit, or in the condition of large misalignment. A forward converter operating in PCM mode limits the current on the primary side. The response of the forward converter was simulated by open- and closed-loop control with LTspice and experimentally verified. Accordingly, the experimental efficiency of the whole system was calculated for full alignment and 3 cm misalignment. The maximum WCS efficiency was 87.52% in the perfectly aligned condition, whereas the maximum efficiency in the 3 cm misalignment was 83.63%. According to the simulation results, these efficiencies are 90.45% and 87.81%, respectively. In addition, the response of the control algorithm was experimentally validated in open-circuit and large misalignments. Accordingly, the developed wireless charging system continues to transmit power under constant load at up to 25% misalignment. In cases of greater misalignment and open circuit, it shuts itself off.

The primary-side controlled WCS proposed in this paper can be used in e-vehicles with similar power and voltage levels using other compensation topologies that are derivatives of the SS topology.

Author Contributions: Conceptualization, E.Y. and M.B.; methodology, E.Y. and M.B.; software, M.B.; validation, M.B.; formal analysis, E.Y.; investigation, E.Y.; resources, E.Y. and M.B.; data curation, M.B.; writing—original draft preparation, E.Y.; writing—review and editing, E.Y.; visualization, E.Y. and M.B.; supervision, E.Y.; project administration, E.Y.; funding acquisition, M.B. All authors have read and agreed to the published version of the manuscript.

Funding: This research and the APC were funded by [Farba Automotive].

Data Availability Statement: Not applicable.

Conflicts of Interest: The authors declare no conflict of interest.

Abbreviations

Parameters

R_o	equivalent load resistance
R_L	load resistance
M	mutual inductance
V_o'	input voltage of rectifier in the secondary side of IPT
V_P	Primary side input voltage of IPT
V_F	output voltage of forward converter
L_P, L_S	primary and secondary side self-inductances of IPT
C_P, C_S	primary and secondary side resonance capacitors of IPT
I_P, I_S	primary and secondary side currents of the IPT
ω	operating angular frequency
Q_P, Q_S	primary and secondary side quality factors of IPT
d	distance between the primary and secondary pads of IPT
N_P, N_S	Turn numbers of primary and secondary coils of IPT
f_0	resonance frequency
η	AC-AC efficiency of IPT
P_o'	input power of the rectifier on the secondary side of IPT
G_{I-V}	current–voltage gain of the IPT
G_{V-V}	voltage–voltage gain of the IPT
η_F	forward converter efficiency
N_F	turn ratio of the forward transformer
D	Duty ratio of the forward converter
L_F, C_F	inductor and capacitor of low-pass filter for forward converter
ω_F	switching frequency of forward converter
$\Delta I_F, \Delta V_F$	the ripple in current and voltage of the forward converter output
k	magnetic coupling
V_o, I_o	load voltage and current
V_{F_test}	forward converter test voltage to control IPT load
I_{F_max1}	maximum forward converter current level to set for CC mode
I_{F_max2}	maximum forward converter current level to set for CV mode

Abbreviations

IPT	inductive power transfer
CC	constant current
CV	constant voltage
PTE	power transfer efficiency of IPT
ESR	effective series resistance
PCM	peak current mode
SOC	state of charge
WCS	wireless charging systems

References

1. Gbey, E.; Turkson, R.F.; Lee, S. A Bibliometric Survey of Research Output on Wireless Charging for Electric Vehicles. *World Electr. Veh. J.* **2022**, *13*, 37. [CrossRef]
2. Li, S.; Li, W.; Deng, J.; Nguyen, T.D.; Mi, C.C. A Double-sided LCC compensation network and its tuning method for wireless power transfer. *IEEE Trans. Veh. Technol.* **2015**, *64*, 2261–2273. [CrossRef]
3. Lu, F.; Zhang, H.; Hofmann, H.; Mi, C.C. An inductive and capacitive integrated coupler and its LCL compensation circuit design for wireless power transfer. *IEEE Trans. Ind. Appl.* **2017**, *53*, 4903–4913. [CrossRef]
4. Song, K.; Li, Z.; Jiang, J.; Zhu, C. Constant current/voltage charging operation for series–series and series–parallel compensated wireless power transfer systems employing primary-side controller. *IEEE Trans. Power Electron.* **2018**, *33*, 8065–8080. [CrossRef]
5. Samanta, S.; Rathore, A.K. A New Inductive Power Transfer Topology Using Direct AC–AC Converter with Active Source Current Waveshaping. *IEEE Trans. Power Electron.* **2018**, *33*, 5565–5577. [CrossRef]
6. Vardani, B.; Tummuru, N.R. A Single-Stage Bidirectional Inductive Power Transfer System with Closed-Loop Current Control Strategy. *IEEE Trans. Transp. Electrif.* **2020**, *6*, 948–957. [CrossRef]
7. Huynh, P.S.; Ronanki, D.; Vincent, D.; Williamson, S.S. Overview and Comparative Assessment of Single-Phase Power Converter Topologies of Inductive Wireless Charging Systems. *Energies* **2020**, *13*, 2150. [CrossRef]
8. Wang, Z.; Wei, X.; Dai, H. Design and Control of a 3 kW Wireless Power Transfer System for Electric Vehicles. *Energies* **2016**, *9*, 10. [CrossRef]
9. Hata, K.; Imura, T.; Hori, Y. Maximum efficiency control of wireless power transfer systems with Half Active Rectifier based on primary current measurement. In Proceedings of the IEEE 3rd International Future Energy Electronics Conference and ECCE Asia (IFEEC 2017-ECCE Asia), Kaohsiung, Taiwan, 3–7 June 2017; pp. 1–6.
10. Corti, F.; Reatti, A.; Pierini, M.; Barbieri, R.; Berzi, L.; Nepote, A.; Pierre, P.D.L. A low-cost secondary-side controlled electric vehicle wireless charging system using a full-active rectifier. In Proceedings of the International Conference of Electrical and Electronic Technologies for Automotive, Milan, Italy, 9–11 July 2018; pp. 1–6.
11. Triviño, A.; González, J.M.; Aguado, J.A. Wireless Power Transfer Technologies Applied to Electric Vehicles: A Review. *Energies* **2021**, *14*, 1547. [CrossRef]
12. Miller, M.; Onar, O.C.; Chinthavali, M. Primary-side power flow control of wireless power transfer for electric vehicle charging. *IEEE J. Emerg. Sel. Top. Power Electron.* **2015**, *3*, 147–162. [CrossRef]
13. Zhao, Q.; Wang, A.; Liu, J.; Wang, X. The Load Estimation and Power Tracking Integrated Control Strategy for Dual-Sides Controlled LCC Compensated Wireless Charging System. *IEEE Access* **2019**, *7*, 75749–75761. [CrossRef]
14. Guo, Y.; Zhang, Y. Secondary Side Voltage and Current Estimation of Wireless Power Transfer Systems. *IEEE Trans. Ind. Appl.* **2022**, *58*, 1222–1230. [CrossRef]
15. Liu, Y.; Madawala, U.K.; Mai, R.; He, Z. Primary-Side Parameter Estimation Method for Bidirectional Inductive Power Transfer Systems. *IEEE Trans. Power Electron.* **2021**, *36*, 68–72. [CrossRef]
16. Gati, E.; Kampitsis, G.; Manias, S. Variable frequency controller for inductive power transfer in dynamic conditions. *IEEE Trans. Power Electron.* **2017**, *32*, 1684–1696. [CrossRef]
17. Li, Z.; Song, K.; Jiang, J.; Zhu, C. Constant current charging and maximum efficiency tracking control scheme for supercapacitor wireless charging. *IEEE Trans. Power Electron.* **2018**, *33*, 9088–9100. [CrossRef]
18. Li, Z.; Wei, G.; Dong, S.; Song, K.; Zhu, C. Constant current/voltage charging for the inductor–capacitor–inductor-series compensated wireless power transfer systems using primary-side electrical information. *IET Power Electron.* **2018**, *11*, 2302–2310. [CrossRef]
19. Li, M.; Deng, J.; Chen, D.; Wang, W.; Wang, Z.; Li, Y. A Control Strategy for ZVS Realization in LCC-S Compensated WPT System with Semi Bridgeless Active Rectifier for Wireless EV Charging. In Proceedings of the 2021 IEEE Energy Conversion Congress and Exposition (ECCE), Vancouver, BC, Canada, 10–14 October 2021; pp. 5823–5827.
20. Wang, Y.; Mostafa, A.; Zhang, H.; Mei, Y.; Zhu, C.; Lu, F. Sensitivity Investigation and Mitigation on Power and Efficiency to Resonant Parameters in an LCC Network for Inductive Power Transfer. *IEEE J. Emerg. Sel. Top. Ind. Electron.* **2021**. [CrossRef]
21. Zhang, Y.; Wang, L.; Guo, Y.; Bo, Q.; Liu, Z. An Optimization Method of Dual-side LCC Compensation Networks Simultaneously Considering Output Power and Transmission Efficiency in Two-directions for BWPT Systems. *IEEE J. Emerg. Sel. Top. Ind. Electron.* **2022**. [CrossRef]
22. Bosshard, R.; Badstübner, U.; Kolar, J.W.; Stevanović, I. Comparative evaluation of control methods for Inductive Power Transfer. In Proceedings of the International Conference on Renewable Energy Research and Applications (ICRERA), Nagasaki, Japan, 11–14 November 2012; pp. 1–6.
23. Liu, F.; Lei, W.; Wang, T.; Nie, C.; Wang, Y. A phase-shift soft-switching control strategy for dual active wireless power transfer system. In Proceedings of the IEEE Energy Conversion Congress and Exposition (ECCE), Cincinati, OH, USA, 1–5 October 2017; pp. 2573–2578.
24. Patil, D.; Sirico, M.; Gu, L.; Fahimi, B. Maximum efficiency tracking in wireless power transfer for battery charger: Phase shift and frequency control. In Proceedings of the IEEE Energy Conversion Congress and Exposition (ECCE), Milwaukee, WI, USA, 18–22 September 2016; pp. 1–8.

25. Pellitteri, F.; Campagna, N.; Castiglia, V.; Damiano, A.; Miceli, R. Design, implementation and experimental results of a wireless charger for E-bikes. In Proceedings of the International Conference on Clean Electrical Power (ICCEP), Otranto, Italy, 2–4 July 2019; pp. 364–369.
26. Joseph, P.K.; Elangovan, D.; Arunkumar, G. Linear control of wireless charging for electric bicycles. *Appl. Energy* **2019**, *255*, 113898. [CrossRef]
27. Pellitteri, F.; Boscaino, V.; Di Tommaso, A.O.; Miceli, R. Wireless battery charging: E-bike application. In Proceedings of the International Conference on Renewable Energy Research and Application (ICRERA), Madrid, Spain, 20–23 October 2013; pp. 247–251.
28. Pellitteri, F.; Boscaino, V.; Miceli, R.; Madawala, U.K. Power tracking with maximum efficiency for wireless charging of E-bikes. In Proceedings of the IEEE International Electric Vehicle Conference (IEVC), Florence, Italy, 17–19 December 2014; pp. 1–7.
29. Hou, J.; Chen, Q.; Zhang, L.; Xu, L.; Wong, S.C.; Chi, K.T. Compact Capacitive Compensation for Adjustable Load-Independent Output and Zero-Phase-Angle Input for High Efficiency IPT Systems. *IEEE J. Emerg. Sel. Top. Power Electron.* **2022**. [CrossRef]
30. Cabrera, A.T.; González, J.M.; Aguado, J.A. Design and Implementation of a Cost-Effective Wireless Charger for an Electric Bicycle. *IEEE Access* **2021**, *9*, 85277–85288. [CrossRef]
31. Skorvaga, J.; Frivaldsky, M.; Pavelek, M. Design of a Wireless Charging System for e-Scooter. *Elektron. Ir Elektrotechnika* **2021**, *27*, 40–48. [CrossRef]
32. Stielau, O.H.; Covic, G.A. Design of loosely coupled inductive power transfer systems. In Proceedings of the International Conference on Power System Technology, Perth, WA, Australia, 4–7 December 2000; pp. 85–90.
33. Yildiriz, E.; Kemer, S.B.; Bayraktar, M. IPT design with optimal use of spiral rectangular coils for wireless charging of e-tricycle scooters. *Eng. Sci. Technol. Int. J.* **2022**, *33*, 101082. [CrossRef]
34. Boys, J.T.; Covic, G.A.; Green, A.W. Stability and control of inductively coupled power transfer systems. *Electr. Power Appl.* **2000**, *147*, 37–43. [CrossRef]
35. Sallán, J.; Villa, J.L.; Llombart, A.; Sanz, J.F. Optimal design of ICPT systems applied to electric vehicle battery charge. *IEEE Trans. Ind. Electron.* **2009**, *56*, 2140–2149. [CrossRef]

A Proposed Controllable Crowbar for a Brushless Doubly-Fed Reluctance Generator, a Grid-Integrated Wind Turbine

Mahmoud Rihan [1], Mahmoud Nasrallah [1], Barkat Hasanin [2] and Adel El-Shahat [3,*]

1 Electrical Engineering Department, Faculty of Engineering, South Valley University, Qena 83521, Egypt; mahmoudrihan@eng.svu.edu.eg (M.R.); nasrallah_1@yahoo.com (M.N.)
2 Electrical Engineering Department, Qena Faculty of Engineering, Al-Azhar University, Qena 83513, Egypt; barkathasanin@yahoo.com
3 Energy Technology Program, School of Engineering, Purdue University, West Lafayette, IN 47906, USA
* Correspondence: asayedah@purdue.edu

Abstract: Brushless doubly fed reluctance generators (BDFRGs) are hopeful generators for using inside variable speed wind turbines (VSWTs), as these generators introduce a promising economical value because of their lower manufacturing and maintenance costs besides their higher reliability. For integrating WT generators, global networks codes require enabling these generators to stay connected under grid disturbances. The behavior of the BDFRG is strongly affected by grid disturbances, due to the small rating of the used partial power converters, as these converters cannot withstand high faults currents which leads to quick tripping of BDFRG. VSWTs can be safeguarded against faults using the crowbar. Usually, the conventual crowbar is shunt connected across the converter to protect it, but this configuration leads to absorbing reactive power with huge amounts from the grid, leading for more voltage decaying and more power system stability deterioration. This study proposes a simpler self-controllable crowbar to enhance the ability of the BDFRG to remain in service under faults. The operation technique of the proposed crowbar is compared to other crowbar operation techniques, the effectiveness of the proposed system would be analyzed. Through the simulation results and behavior analysis, the proposed crowbar technique demonstrates a decent improvement in the conduct of the studied system under faults.

Keywords: brushless doubly fed reluctance generator (BDFRG); crowbar; symmetrical fault; unsymmetrical fault; wind turbine (WT)

1. Introduction

Because of the ongoing rise in the prices of fossil resources, which mostly have an exhausted impact on the world's economies, as well as the harmful influence of these resources on the environmental footprint, the world rapidly resorts to utilizing sustainable resources. Indeed, the wind energy conversion system (WECS) has emerged as one of the fastest-growing energy systems [1–4]. Doubly fed induction generator (DFIGs) introduced a main important advantage since these generators are based on using a partial converter which means lower cost than the generators that use a complete power converter, leading to a great spread of DFIGs in wind energy markets. Brushless doubly fed machines (BDFMs) are slip recovery machines that reduce the required power converter if the required speed control range is limited, resulting in a significant cost saving for these machines [5]. As a result, using this BDFM is a cost-effective option for using inside the variable-speed WECS. Another main advantage of the BDFMs over the traditional DFIGs is their reliability, as the brush gear and slip rings were removed. Nowadays, both the brushless doubly fed reluctance machine (BDFRM) and brushless doubly fed induction machine (BDFIM) attract the most attention from researchers, especially on the subjects of stability, maximizing the extracted power and power quality. The main significances of the BDFRM's rotor design are to give it; higher reliability, robustness, and operation free from maintenance.

Moreover, considering the lack of rotor copper losses, the BDFRM is expected to outperform the BDFIM in terms of efficiency [6–8]. As a result, the brushless doubly fed reluctance generator (BDFRG) is discovered to be the most appealing for WECS [9–11]. Moreover, the BDFRG has been studied in a lot of fields by researchers as mentioned in papers [12–15].

Some of the grid codes considered that WTs have to be able to remain connected under zero voltage for a duration time up to 150 ms [16–20]. As the stator windings of the BDFRG are tied directly to the network, any occurrence of grid disturbances especially voltage dip can easily lead to an abrupt absence of the BDFRG magnetization, raising the current values in the machine side converter (MSC) above the threshold value and also raising the voltage of the DC-link above the threshold value, which can easily lead to destroying the power converter [21]. So, there are many different technical challenges facing the ability of the BDFRGs to satisfy the grid codes' requirements of keeping these generators connected under faults, principally due to the lower ratings of the used partial power converters.

This study is the first one to boost the ability of the BDFRG of staying connected during faults by using the crowbar. Commonly, currently, there are two used solutions to promote the ability of the doubly fed generators to keep connected during faults: using the crowbar and applying the demagnetization method [20]. Usually, when the crowbar was applied to protect the doubly fed generators during the occurrence of faults, the Machine Side Converter (MSC) is short-circuited, absorbing from the grid large amounts of reactive power, meaning more voltage dropping and more instability for the grid [22]. While on the other side, applying the demagnetization method [23,24], based on controlling the output of the MSC for tracing and counteracting the stator flux oscillations to eliminate the occurred transients on the induced electromagnetic force in the rotor winding. However, the industrial realization of the demagnetization method is very complex [20].

This paper introduces a proposed solution for increasing the ability of the grid-connected BDFRG WT to remain in service under the occurrence of faults and subsequently increasing the ability of the BDFRG to satisfy the grid code requirements. The proposed solution is mainly depending on using a new automatically controllable crowbar protection technique. One of the main aims of the proposed solution is keeping the connecting of the BDFRG MSC during the faults, reducing reactive power absorption from the network. For assessing the proposed solution efficacy, the "BDFRG grid-connected wind farm" performance would be examined under the occurrence of heavy conditions of different faults with and without using the proposed solution.

2. BDFRG Dynamic Model

The BDFRG dynamic model is based on the theory of space vector, shown in Equations (1)–(3) [4]. The process of decoupling both the torque-producing and flux-producing current components is only possible in a special kind of frame transformation. This is mainly because of the bizarre structure of the BDFRG and the fact that the primary and secondary windings electrical quantities (current, flux, etc.) both have variant pole numbers and also different frequencies. The primary winding quantities are transformed into a general reference frame that rotates at a speed of ω while the secondary winding quantities are transformed into another frame that rotates at speed of $(\omega_r - \omega)$ [25]. The general frame speed is preferred to be the same as the primary winding supply frequency. The resulting dynamic model based on the dual frame transformation is described by Equation (1). The flux equation is presented in Equation (2). Finally, the electromagnetic torque and mechanical speed expressions can be stated as in Equation (3) [26,27].

$$v_{dp} = r_p \, i_{dp} + \frac{d}{dt} \lambda_{dp} - \omega \, \lambda_{qp}$$

$$v_{qp} = r_p \, i_{qp} + \frac{d}{dt} \lambda_{qp} + \omega \, \lambda_{dp}$$

$$v_{dc} = r_c \, i_{dc} + \frac{d}{dt} \lambda_{dc} - (\omega_r - \omega) \, \lambda_{qc}$$

$$v_{qc} = r_c\, i_{qc} + \frac{d}{dt}\lambda_{qc} + (\omega_r - \omega)\,\lambda_{dc} \tag{1}$$

$$\begin{aligned}
\lambda_{dp} &= L_p\, i_{dp} + L_{pc}\, i_{dc} \\
\lambda_{qp} &= L_p\, i_{qp} - L_{pc}\, i_{qc} \\
\lambda_{dc} &= L_c\, i_{dc} + L_{pc}\, i_{dp} \\
\lambda_{qc} &= L_c\, i_{qc} - L_{pc}\, i_{qp}
\end{aligned} \tag{2}$$

$$T_e = \frac{3}{2}\, p_r \left[\lambda_{dp}\, i_{qp} - \lambda_{qp}\, i_{dp}\right]$$

$$\left(J_r + n_g^2\, J_g\right)\frac{d\omega_{rm}}{dt} = n_g\, T_m + n_g^2\, T_e \tag{3}$$

As described in Figure 1, the system consists of the BDFRG driven by WT The BDFRG is tied to the network by the power and control windings. The power winding is directly connected, while the control winding is connected via two converters: Machine Side Converter (MSC) and Grid Side Converter (GSC). A capacitance is placed between two converters. Each converter has its own controller in order to assure that it functions properly. Both the MSC and GSC controllers are shown in Figure 1. The MSC controller has two branches; one of them is connected to MPPT, this branch calculates the reference quadrature secondary voltage by using indirect field-oriented control. The other branch calculates the reference direct secondary voltage. These two reference voltages are transformed from dq frame to abc frame by using (θ_s) to obtain appropriate gate signals from Sinusoidal Pulse Width Modulation (SPWM). On the other hand, the GSC controller has also two branches as illustrated in Figure 1. The two branches calculate reference quadrature and direct grid voltages. Then, they are transformed from dq frame to abc frame by using (θ_g) to get appropriate gate signals from SPWM to adjust the DC link voltage and secondary reactive power on its reference values.

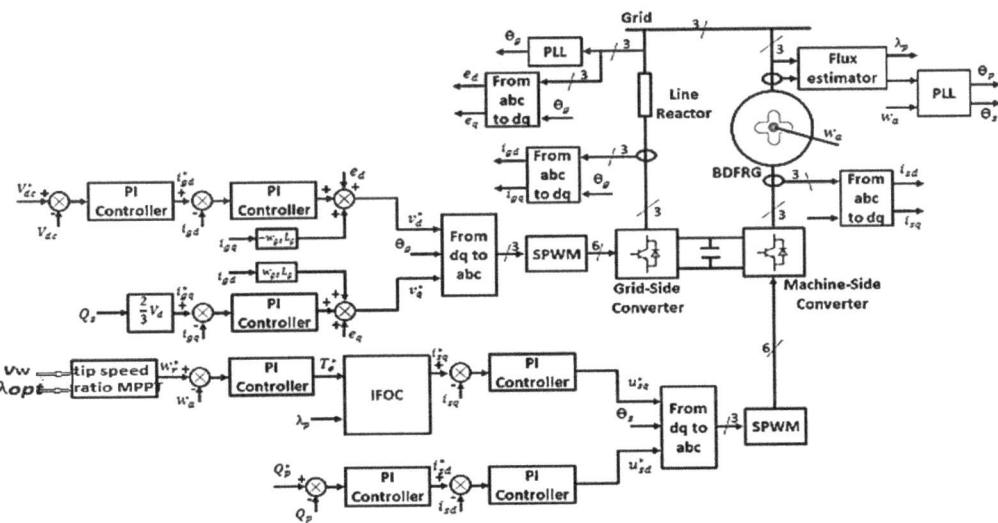

Figure 1. Block diagram of the studied BDFRG.

3. The Crowbar

The crowbar is one of the used solutions in the cases of severe disturbances such as heavy faults. In general, the crowbar provides a safe path for the heavy fault currents by short-circuiting the original path, subsequently damping the destroying fault currents and protecting the generator from the increased (overrated) currents. Basically, crowbar consists of a three-phase resistance called crowbar resistance. These resistances could be connected and disconnected by means of power electronics switches. Severe disturbances lead to raising the currents in the generator above the allowable values, and also disturb the DC link voltage. In general, the activation (connection) process of the crowbar protection system is carried out only in the case of severe disturbances (heavy faults). After clearing the disturbance (fault), if the current and the DC link voltage are returned back to its allowable values, the crowbar will be deactivated (disconnected). Moreover, the generator returned back to its normal configuration. If the current and the DC link voltage are not returned to its allowable values, the activation of the crowbar protection system can be restarted. Moreover, the crowbar has been studied in a lot of fields by researchers as mentioned in the paper [28–30].

4. The Proposed Crowbar Control Strategy

The crowbar activation and deactivation processes are important issues, there are many used techniques for crowbar activation and deactivation processes. In Ref. [29], an Adaptive Neuro Fuzzy Inference System (ANFIS) is used to produce the crowbar control signal which lead to complicating the crowbar system.

This work aimed to propose a simple crowbar control strategy. As shown in Figure 2, the used crowbar technique is the outer crowbar. The main components of the proposed crowbar control system are voltage measuring units for all phases, a control program and an automatic switch. The main aim of the crowbar control program is producing the control signal which would control the automatic switch. The main idea of the proposed strategy is based on continuously monitoring and measuring the per unit rms terminal voltage for each phase individually. According to the flowchart shown in Figure 3, if the measured rms voltage value, of each phase, is within the predefined voltage constraint limits (more than or equals to 0.7 p.u.), the control scheme would behave in this case as a normal steady-state operating condition, then the control program would set the output control signal to (1). Otherwise, if any one of the measured rms voltage values is lower than the minimum predefined voltage constrain limit "0.7 p.u.", the control scheme would behave with this case as a faulty condition, then the control program would set the output control signal to (0). In the case "Output control signal = 1"; the automatic switch would be switched on and bypassing the crowbar "the crowbar is deactivated", while in the case "Output control signal = 0", the automatic switch would be switched off, activating the crowbar. So, the operation technique of the proposed method is very simple in comparison with the methods that were used in both Ref. [20] and Ref. [29] to achieve the same goal, whereas Ref. [20] used a complex fuzzy control system and improved its performance by adding a PI to the used fuzzy control system. Ref. [29] used a complex ANFIS system.

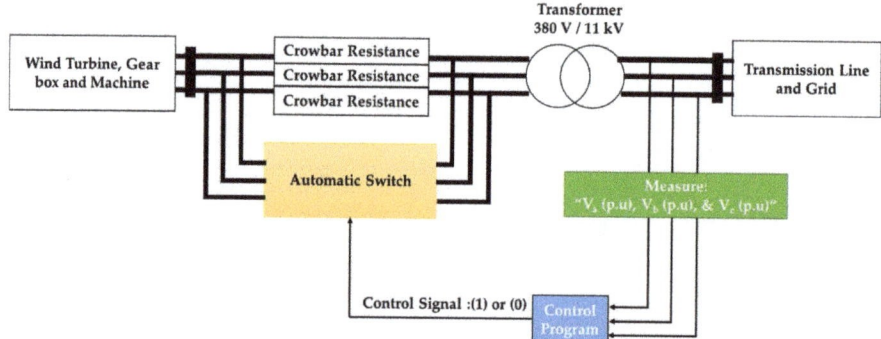

Figure 2. Block diagram of the proposed crowbar technique.

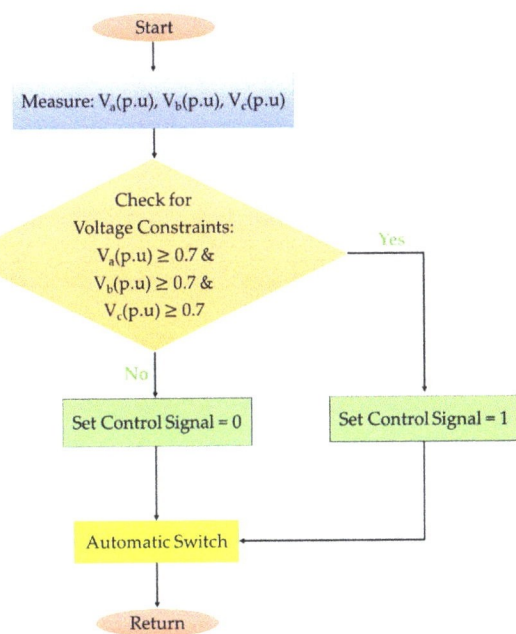

Figure 3. Flow chart of the proposed control for the crowbar.

5. Simulation Results

The studied system is shown in Figure 4, where the BDFRG (supported by the proposed crowbar) is tied to the network by a transmission line after the coupling transformer. The main data of the simulated BDFRG and wind turbine are described in Tables A1 and A2 (Appendix A) [4].

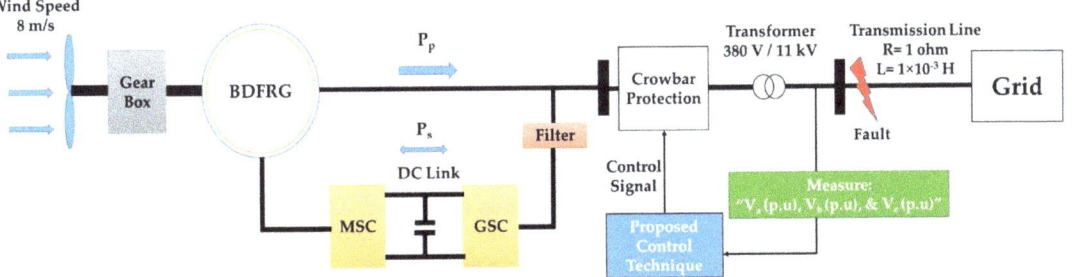

Figure 4. Studied system.

To examine the efficacy of the proposed crowbar control technique, under the occurrence of different heavy disturbances, this work shows and analyses the performance of the studied system without and with using the proposed crowbar control strategy. The studied disturbances are: three-line to ground fault, single line to ground fault, double line fault and double line to ground fault. The disturbances occur at the beginning of the transmission line next to the coupling busbar (11 kV Busbar), applied at the instant of "time = 1 s" for 150 ms duration. According to the used methodology in Ref. [30], the adequate crowbar resistance value for the studied system was 10 times the secondary "control" winding resistance value. To ensure monitoring the total actual performance of the BDFRG wind turbine under the studied faults, all the protection system devices were deactivated. The simulation works implemented by MATLAB/SIMULINK (2013 b).

5.1. Symmetrical Fault (Three Line to Ground Fault)

The per unit rms terminal voltage of the BDFRG, as shown in Figure 5, at the instant of fault occurrence, the terminal voltage dropped to zero p.u. for 150 ms due to the occurrence of the studied three phase to ground fault; then, after fault clearance, the terminal voltage returns to its original value (1 p.u.).

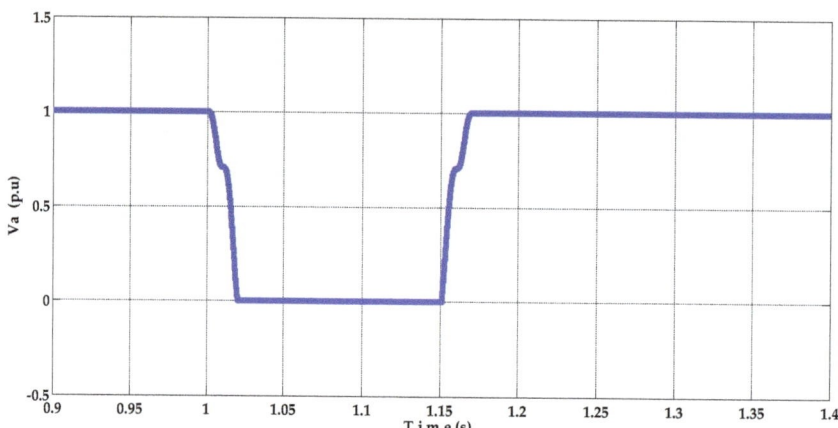

Figure 5. Per unit rms terminal voltage (Va) of the studied wind farm main coupling point under symmetrical fault occurrence.

The active power of the BDFRG (with and without using the proposed crowbar) is shown in Figure 6. As obvious in the case of "without using the proposed crowbar", during the fault, the active power totally dropped to zero kW for 150 ms (fault duration time). In the case of using the proposed crowbar, during the fault, the active power was effectively improved and quickly returned to its pre-fault value.

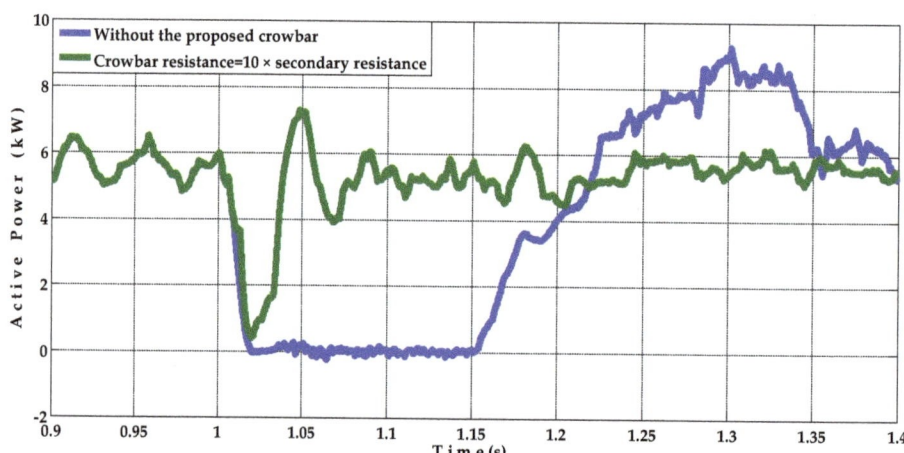

Figure 6. Active power of BDFRG with and without the proposed crowbar under symmetrical fault occurrence.

The reactive power of the BDFRG (with and without the proposed crowbar) is shown in Figure 7. As shown, the reactive power was adjusted at zero value (unity power factor) before the fault occurrence. Following the clearance of the fault, in the case of "without using the proposed crowbar", the absorbed reactive power, by the BDFRG from the grid, reached about 5.04 kvar. While in the case of using the proposed crowbar, after fault clearance, the absorbed reactive power was reduced to 2.165 kvar only and quickly improved until reaching its pre-fault value.

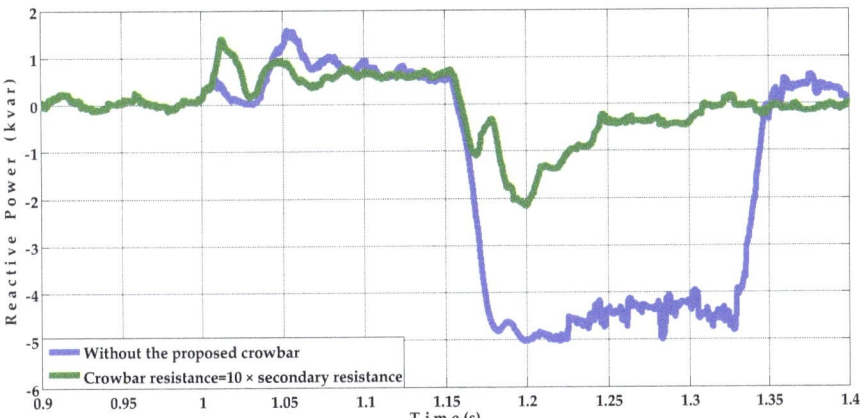

Figure 7. Reactive power of BDFRG with and without the proposed crowbar under symmetrical fault occurrence.

The rotor speed of the BDFRG (with and without using the proposed crowbar) is shown in Figure 8, which has a reference value equal to 1160 rpm. During the fault, in the case of "without using the proposed crowbar", the rotor rapidly accelerated, then after fault clearance, the rotor speed reached about 1464 rpm, which led to a decrease in the power coefficient of the WT from 0.48 to less than 0.3846 as shown in Figure 9. In the case of using the proposed crowbar, the rotor speed increased instantaneously to about 1213 rpm only, then quickly improved.

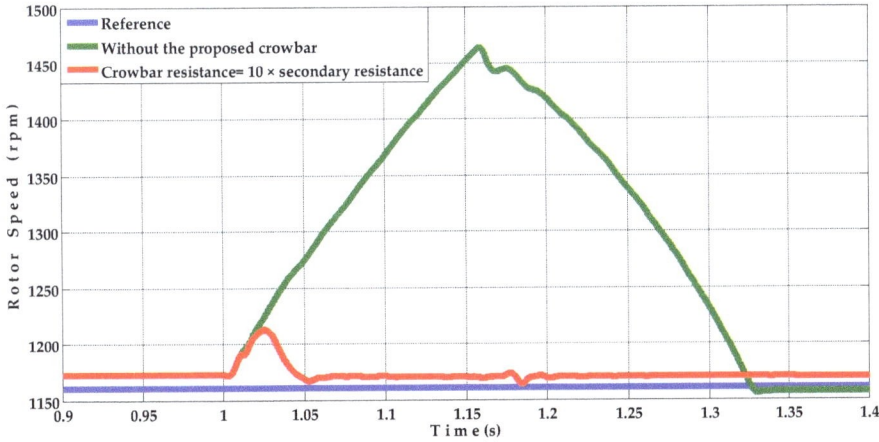

Figure 8. Rotor speed of BDFRG with and without the proposed crowbar under symmetrical fault occurrence.

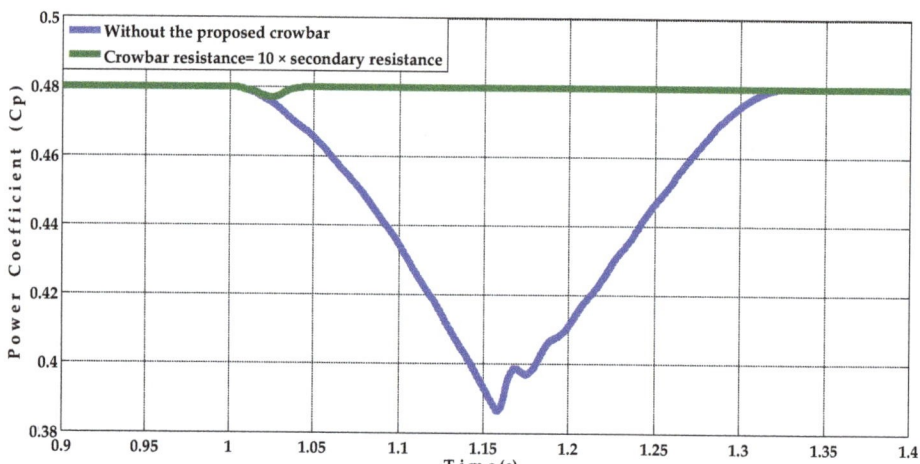

Figure 9. Power coefficient of wind turbine with and without the proposed crowbar under symmetrical fault occurrence.

The primary and secondary currents of the BDFRG (without and with using the proposed crowbar) are shown in Figures 10 and 11 in the same order. In the case of "without using the proposed crowbar", during the fault, both currents were increased for a certain period. After the fault clearance, the primary current was increased to about 212% (19.74 A) of the pre-fault value (9.32 A) and the secondary current was increased to about 216% (28.25 A) of the pre-fault value (13.05 A), while in the case of using the proposed crowbar, as shown in Figures 10b and 11b, after the fault clearance, both the primary and secondary currents were effectively improved. As the primary current increased to (15.25 A), while the secondary current increased to (22.86 A) and then quickly both the primary and secondary currents were effectively improved.

The dc link voltage of the BDFRG (with and without using the proposed crowbar) is shown in Figure 12, which has a reference value equals to 710 V. Under the fault occurrence, in the case of "without using the proposed crowbar", the dc link voltage was decreased to about 499.2 V. In the case of using the proposed crowbar, the dc link voltage decreased instantaneously to about 587.8 V only, then the dc link voltage improved and returned quickly to its pre-fault value.

5.2. Unsymmetrical Fault

5.2.1. Single-Line to Ground Fault (The Fault Is Applied at Phase a)

The per unit rms terminal voltage of the BDFRG, as shown in Figure 13, at the instant of fault occurrence, the terminal voltage dropped to zero p.u. for 150 ms due to the occurrence of the studied single phase to ground fault, then after fault clearance, the terminal voltage returns to its original value (1 p.u.).

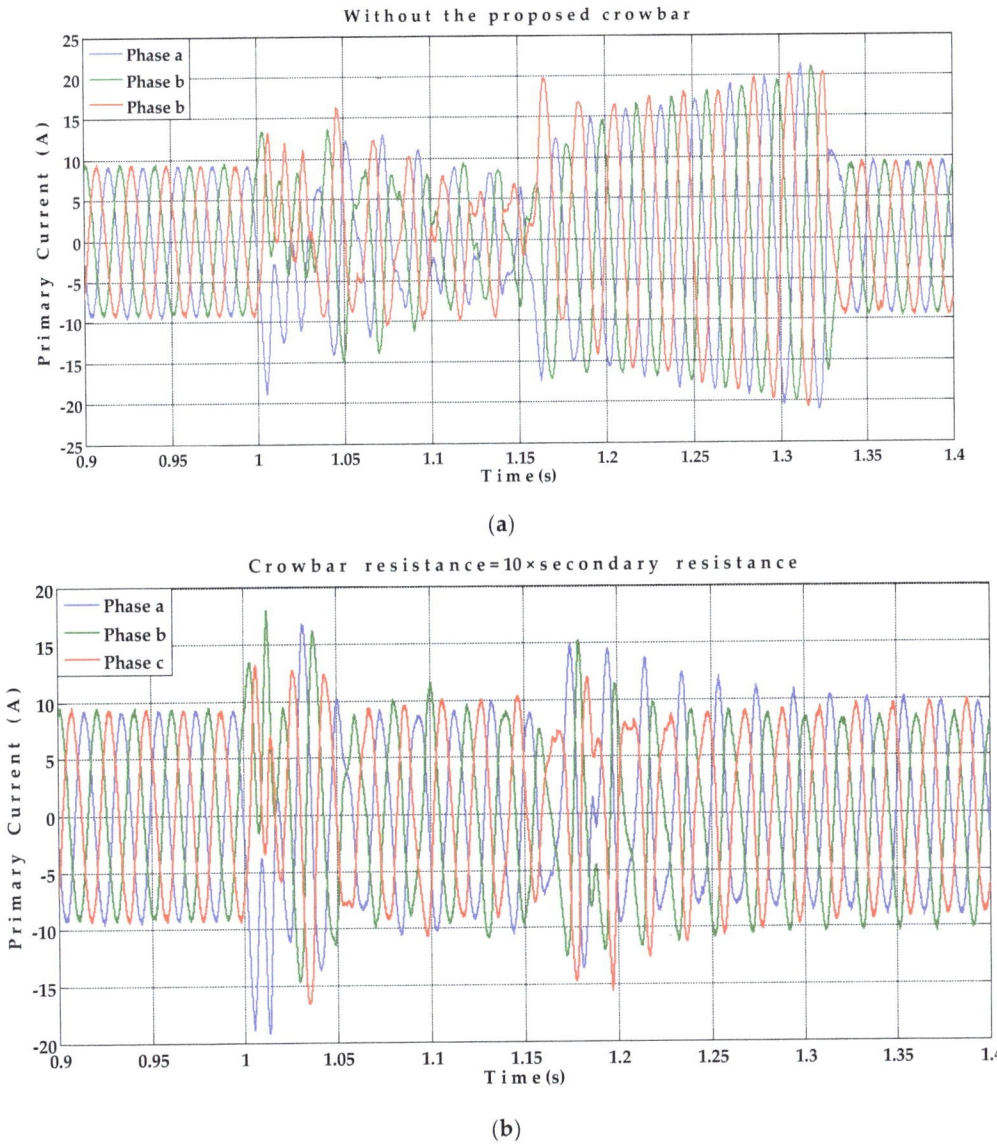

Figure 10. Primary current of BDFRG without and with the proposed crowbar protection under symmetrical fault occurrence: (**a**). Without the proposed crowbar; (**b**). With the proposed crowbar.

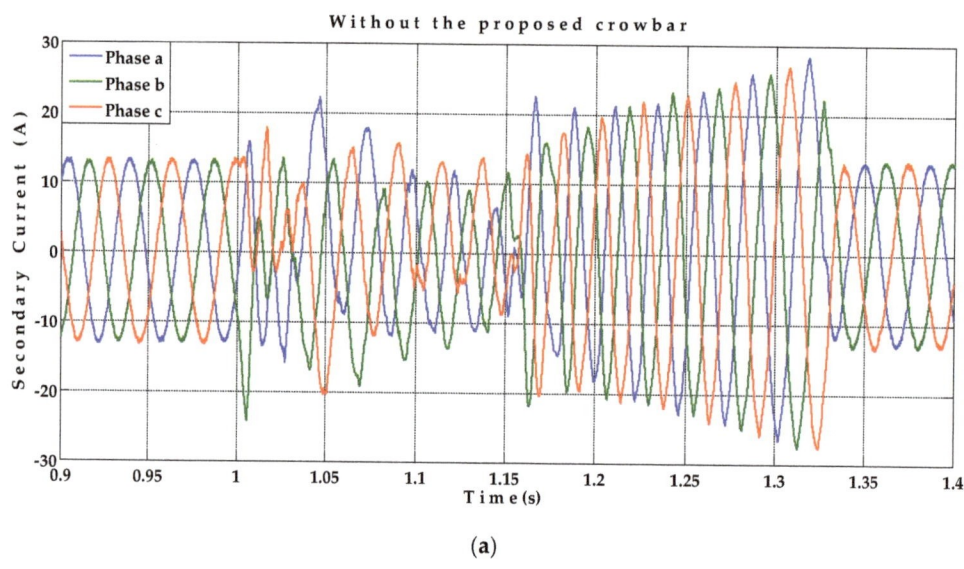

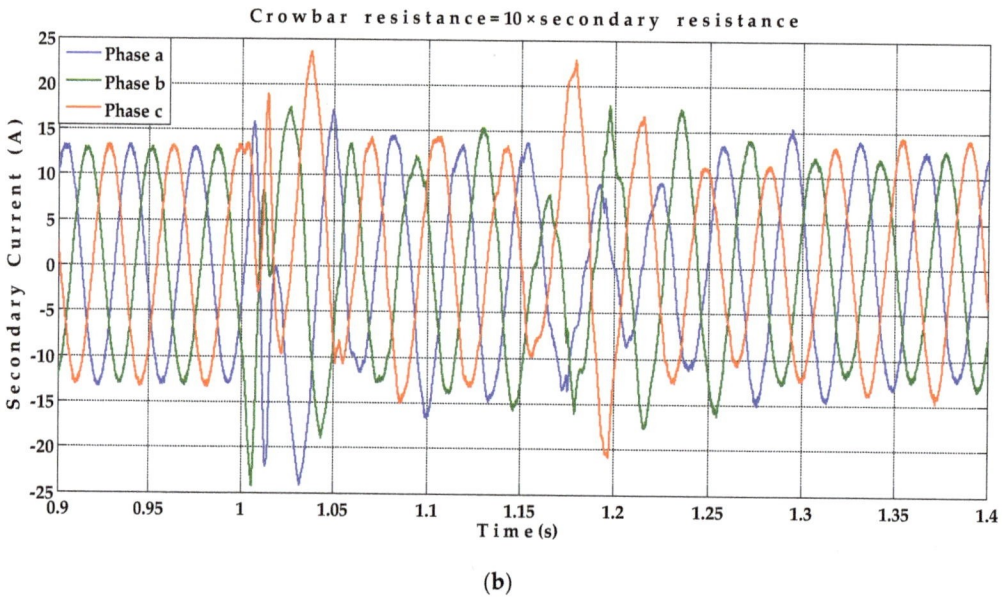

Figure 11. Secondary current of BDFRG without and with the proposed crowbar protection under symmetrical fault occurrence: (**a**) Without the proposed crowbar; (**b**) With the proposed crowbar.

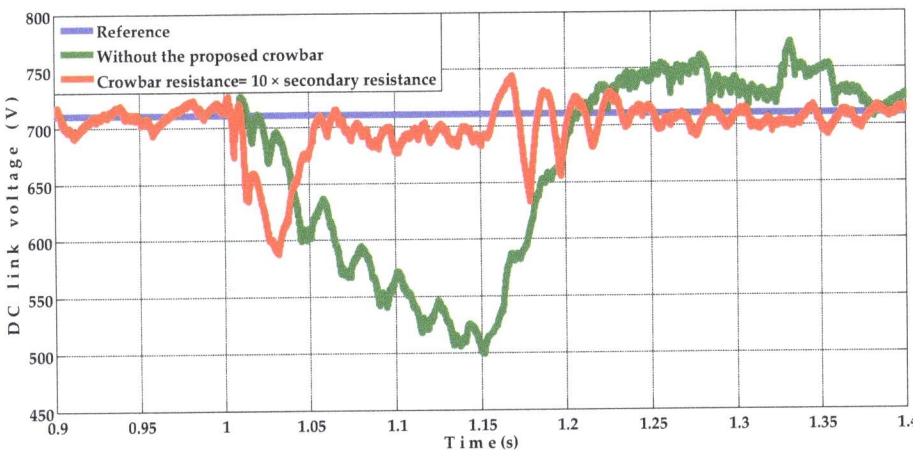

Figure 12. DC link voltage of BDFRG with and without the proposed crowbar under symmetrical fault occurrence.

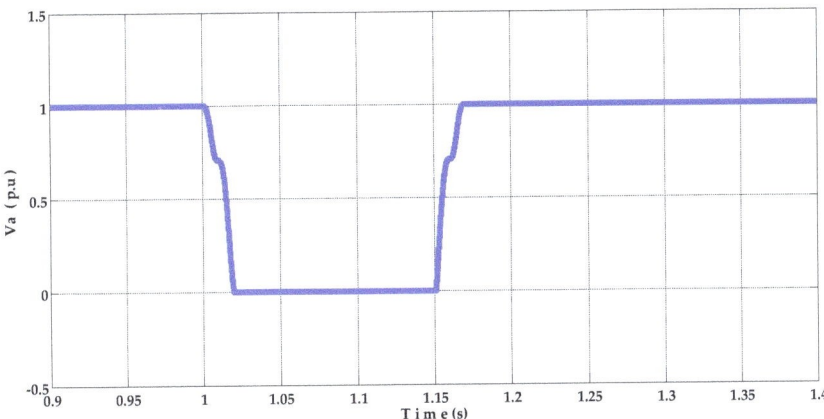

Figure 13. Per unit rms terminal voltage (Va) of the studied wind farm main coupling point under single-line to ground fault occurrence.

The active power of the BDFRG (with and without using the proposed crowbar) is shown in Figure 14. As obvious in the case of "without using the proposed crowbar", during the fault, the active power dropped to 3.78 kW. In the case of using the proposed crowbar, during the fault, the active power was effectively improved and quickly returned to its pre-fault value.

The reactive power of the BDFRG (with and without using the proposed crowbar) is shown in Figure 15. As shown, the reactive power was adjusted at zero value (unity power factor) before the fault occurrence. Following the clearance of the fault, in the case of "without using the proposed crowbar", the absorbed reactive power, by the BDFRG from the grid, reached about 2.115 kvar. In the case of using the proposed crowbar, after fault clearance, the absorbed reactive power was reduced to 1.158 kvar only and quickly improved until reaching its pre-fault value.

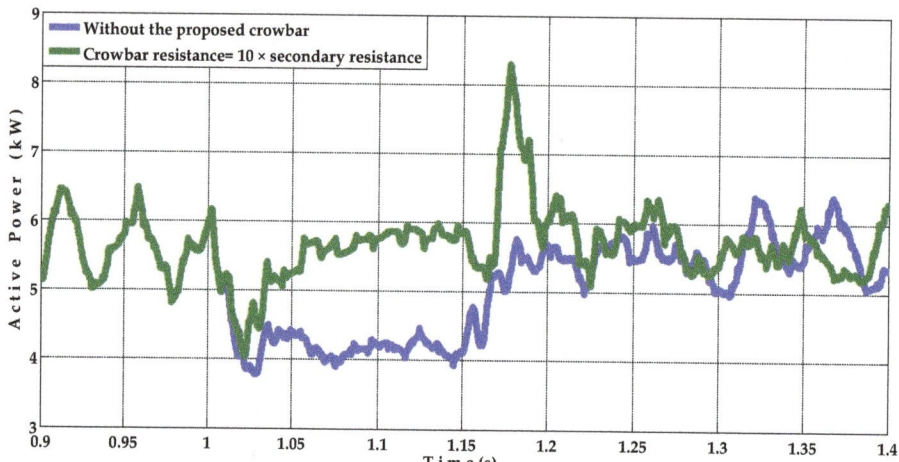

Figure 14. Active power of BDFRG with and without the proposed crowbar under single-line to ground fault occurrence.

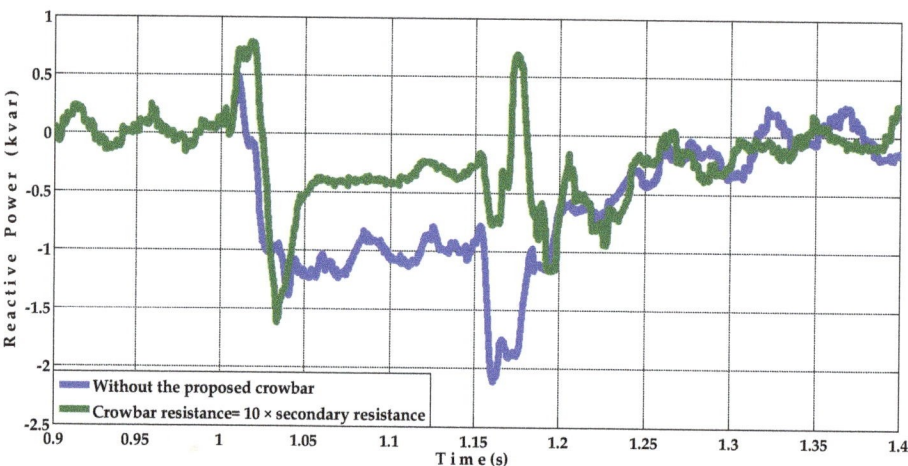

Figure 15. Reactive power of BDFRG with and without the proposed crowbar under single-line to ground fault occurrence.

The rotor speed of the BDFRG (with and without using the proposed crowbar) is shown in Figure 16, which has a reference value equals to 1160 rpm. During fault, in the case of "without using the proposed crowbar", the rotor rapidly accelerated and the rotor speed reached about 1184 rpm, which led to a decrease in the power coefficient of the WT from 0.48 to less than 0.4794 as shown in Figure 17. In the case of using the proposed crowbar, the rotor speed increased instantaneously to about 1186 rpm, but with damped oscillations than in the other case, then quickly improved.

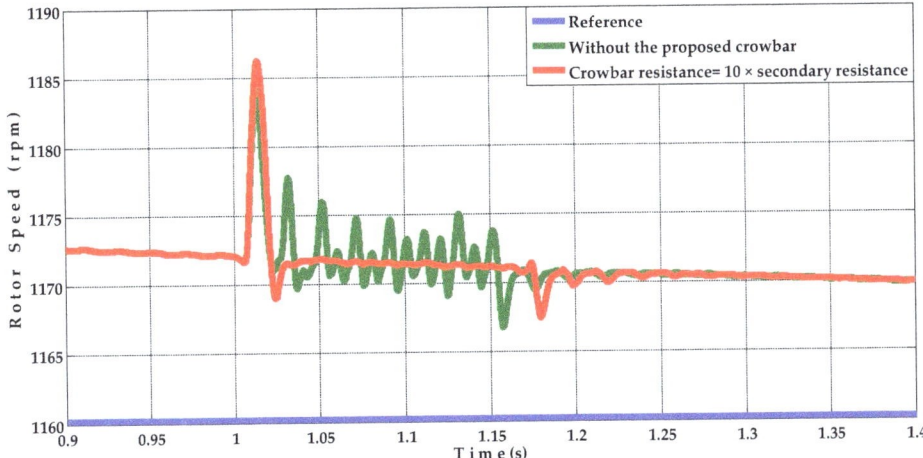

Figure 16. Rotor speed of BDFRG with and without the proposed crowbar under single line to ground fault occurrence.

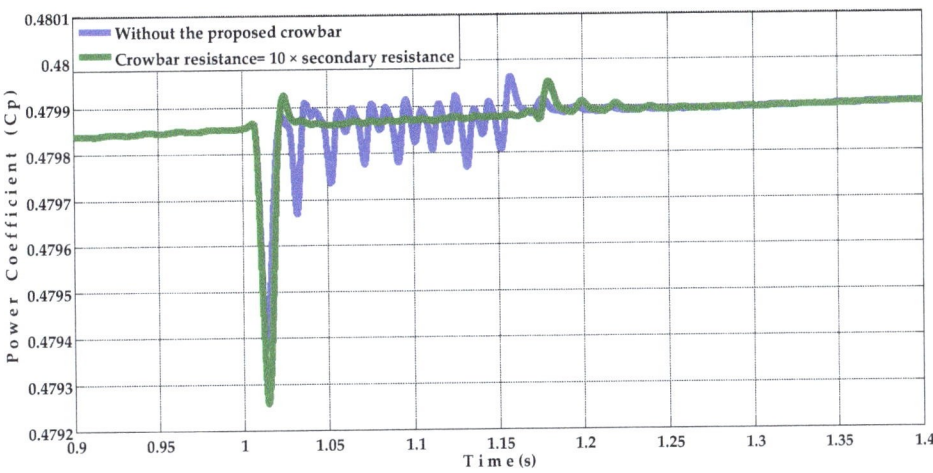

Figure 17. Power coefficient of wind turbine with and without the proposed crowbar under single-line to ground fault.

The primary and secondary currents of the BDFRG (without and with using the proposed crowbar) are shown in Figures 18 and 19 in the same order. In the case of "without using the proposed crowbar", during the fault, both currents were increased, the primary current was increased to about 190% (17.68 A) of the pre-fault value (9.32 A) and the secondary current was increased to about 181% (23.58 A) of the pre-fault value (13.05 A), while in the case of using the proposed crowbar, as shown in Figures 18b and 19b, during the fault, both the primary and secondary currents were effectively improved.

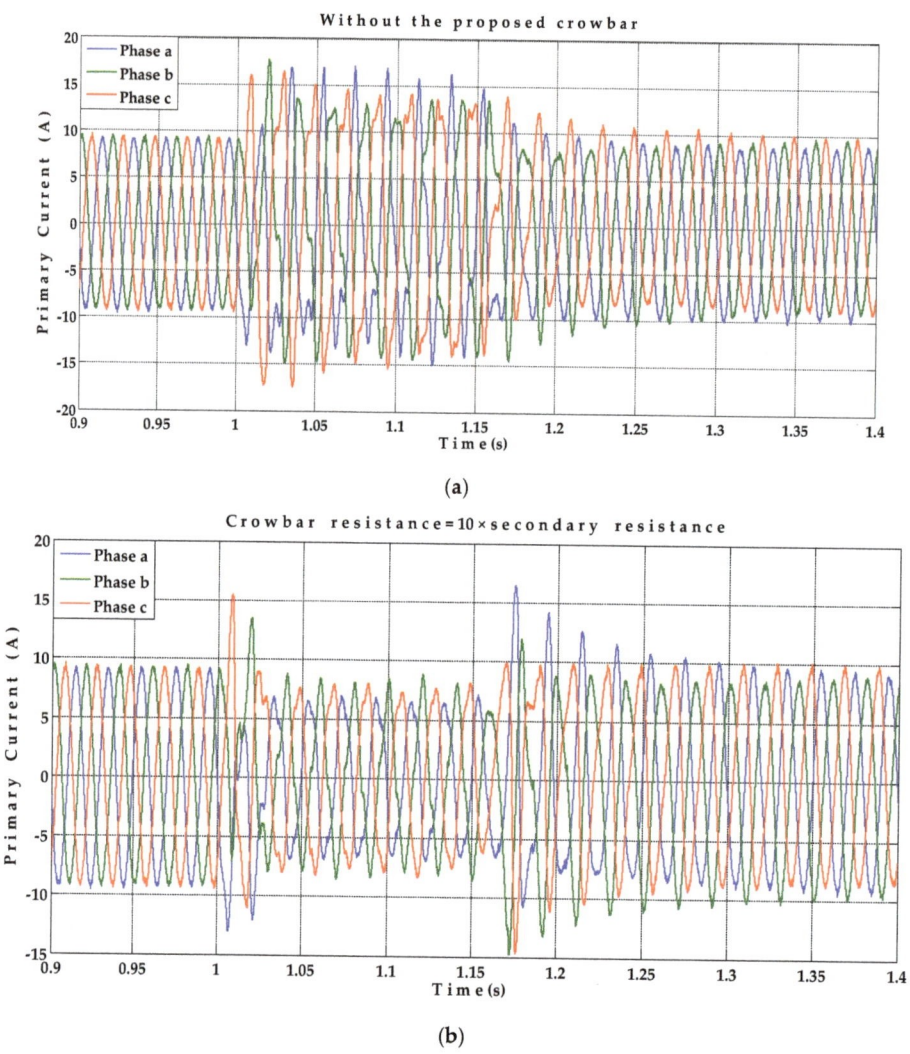

Figure 18. Primary current of BDFRG without and with the proposed crowbar protection under single-line to ground fault: (a) Without the proposed crowbar; (b) With the proposed crowbar.

The dc link voltage of the BDFRG (with and without using the proposed crowbar) is shown in Figure 20, which has a reference value equal to 710 V. Under the fault occurrence, in the case of "without using the proposed crowbar", the dc link voltage was decreased to about 642 V. In the case of using the proposed crowbar, the dc link voltage increased to about 800 V, but with damped oscillations than in the other case, then the dc link voltage improved and returned to its pre-fault value.

5.2.2. Line to Line Fault (The Fault Is Applied at Phases a and b)

The per unit rms terminal voltage of the BDFRG, as shown in Figure 21, at the instant of fault occurrence, the terminal voltage dropped to 0.5 p.u. for 150 ms due to the occurrence of the studied line to line fault, then after fault clearance, the terminal voltage returns to its original value (1 p.u.).

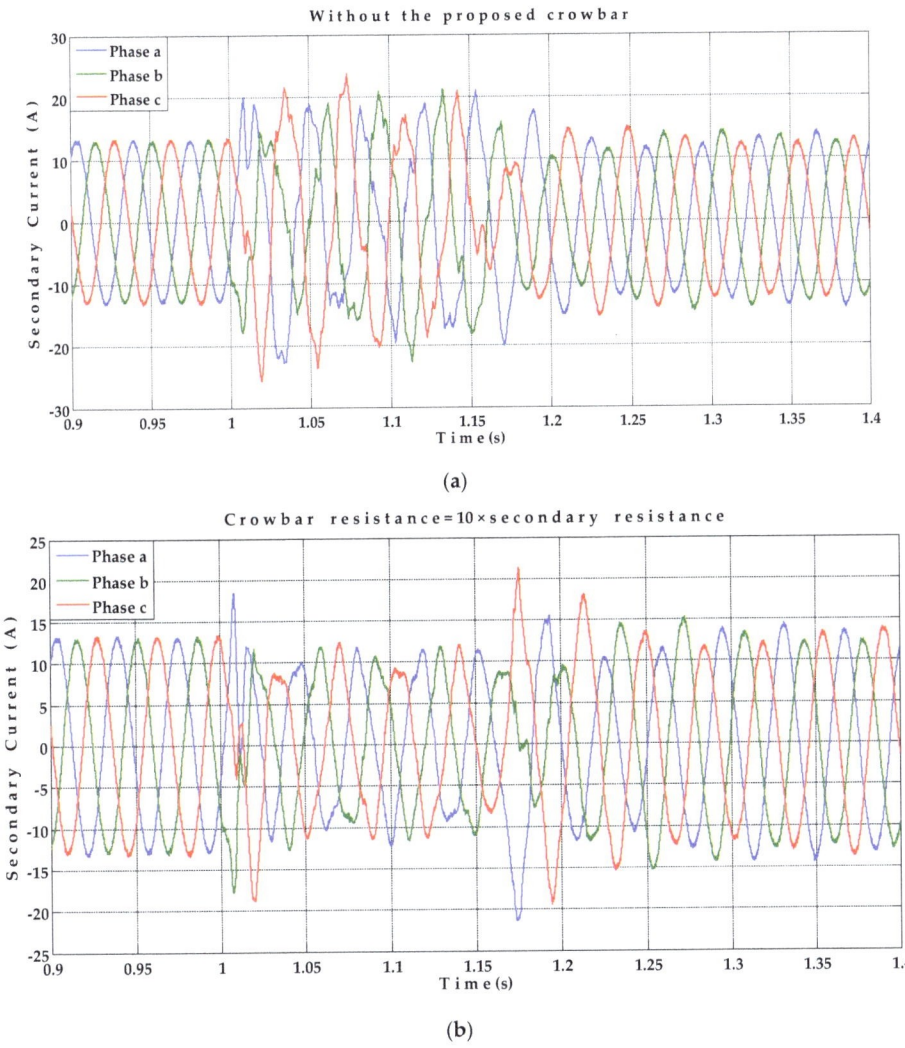

Figure 19. Secondary current of BDFRG without and with the proposed crowbar under single-line to ground fault: (**a**) Without the proposed crowbar; (**b**) With the proposed crowbar.

The active power of the BDFRG (with and without using the proposed crowbar) is shown in Figure 22. As obvious in the case of "without using the proposed crowbar", during the fault, the active power dropped to 1.99 kW. In the case of using the proposed crowbar, during the fault, the active power was effectively improved and quickly returned to its pre-fault value.

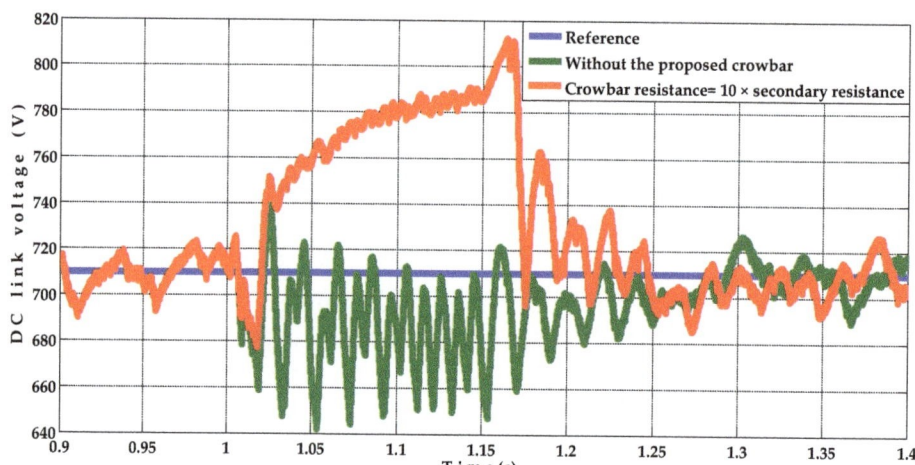

Figure 20. DC link voltage of BDFRG with and without the proposed crowbar under single-line to ground fault occurrence.

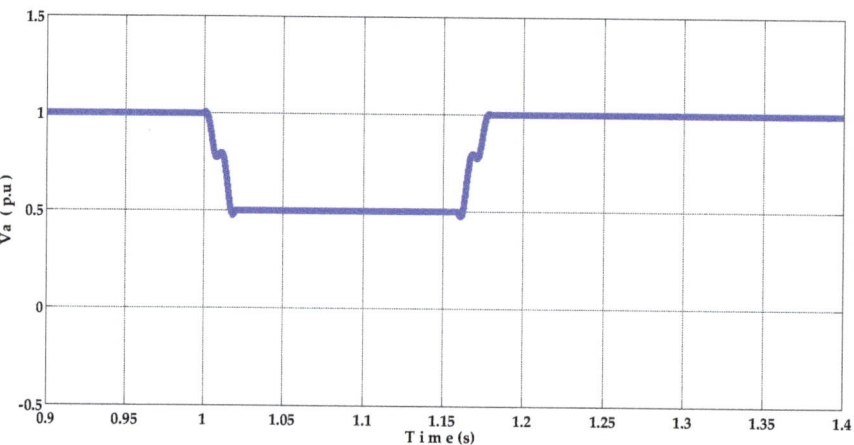

Figure 21. Per unit rms terminal voltage (Va) of the studied wind farm main coupling point under line to line fault occurrence.

The reactive power of the BDFRG (with and without using the proposed crowbar) is shown in Figure 23. As shown, the reactive power was adjusted at zero value (unity power factor) before the fault occurrence. Following the clearance of the fault, in the case of "without using the proposed crowbar", the absorbed reactive power, by the BDFRG from the grid, reached about 3.364 kvar for a certain period. In the case of using the proposed crowbar, after fault clearance, the absorbed reactive power was reduced to 1.325 kvar only and quickly improved until reaching its pre-fault value.

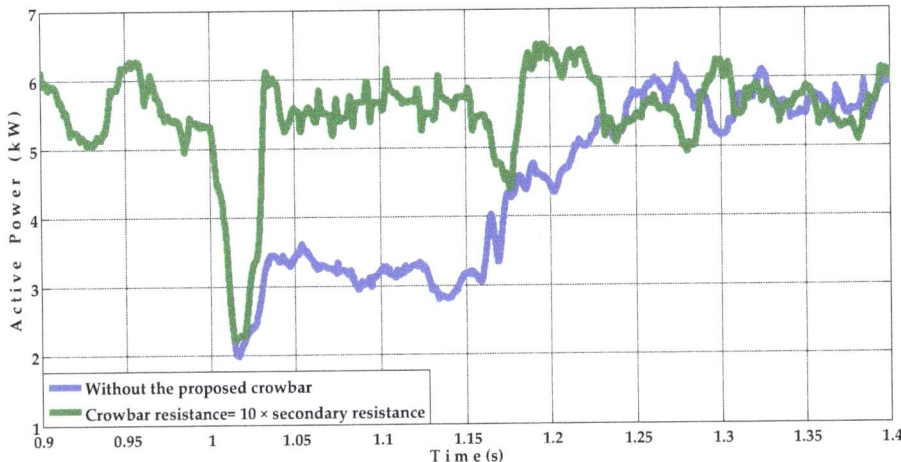

Figure 22. Active power of BDFRG with and without the proposed crowbar under line to line fault occurrence.

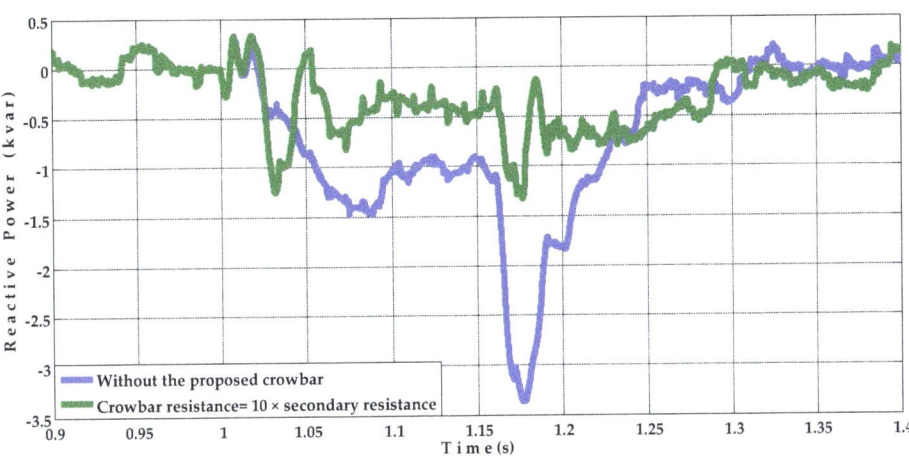

Figure 23. Reactive power of BDFRG with and without the proposed crowbar under line-to-line fault occurrence.

The rotor speed of the BDFRG (with and without using the proposed crowbar) is shown in Figure 24, which has a reference value equal to 1160 rpm. During the fault, in the case of "without using the proposed crowbar", the rotor rapidly accelerated and the rotor speed reached about 1196 rpm, which led to a decrease in the power coefficient of the WT from 0.48 to less than 0.4787 as shown in Figure 25. In the case of using the proposed crowbar, the rotor speed increased instantaneously to about 1198 rpm, but with damped oscillations than in the other case, then quickly improved.

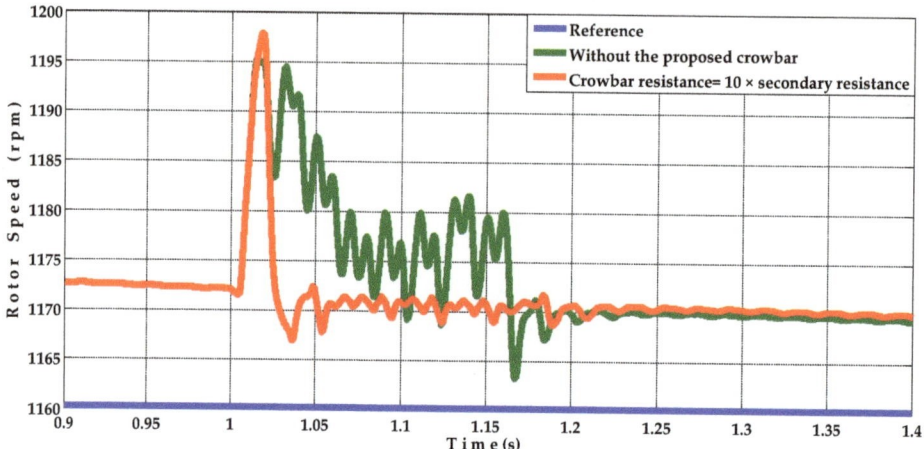

Figure 24. Rotor speed of BDFRG with and without the proposed crowbar under line to line fault occurrence.

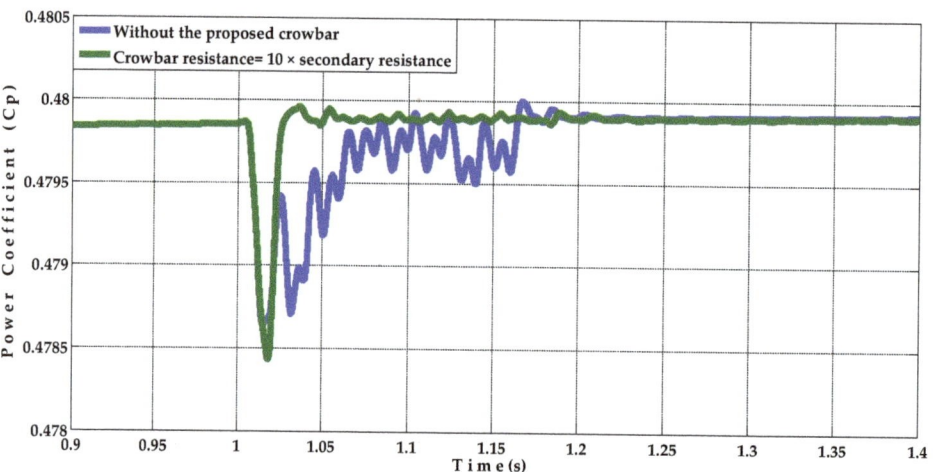

Figure 25. Power coefficient of wind turbine with and without the proposed crowbar under line to line fault occurrence.

The primary and secondary currents of the BDFRG (without and with using the proposed crowbar) are shown in Figures 26 and 27 in the same order. In the case of "without using the proposed crowbar", during the fault, both currents were increased; the primary current was increased to about 220% (20.47 A) of the pre-fault value (9.32 A) and the secondary current was increased to about 215% (28.04 A) of the pre-fault value (13.05 A), while in the case of using the proposed crowbar, as shown in Figures 26b and 27b, during the fault, both the primary and secondary currents were effectively improved.

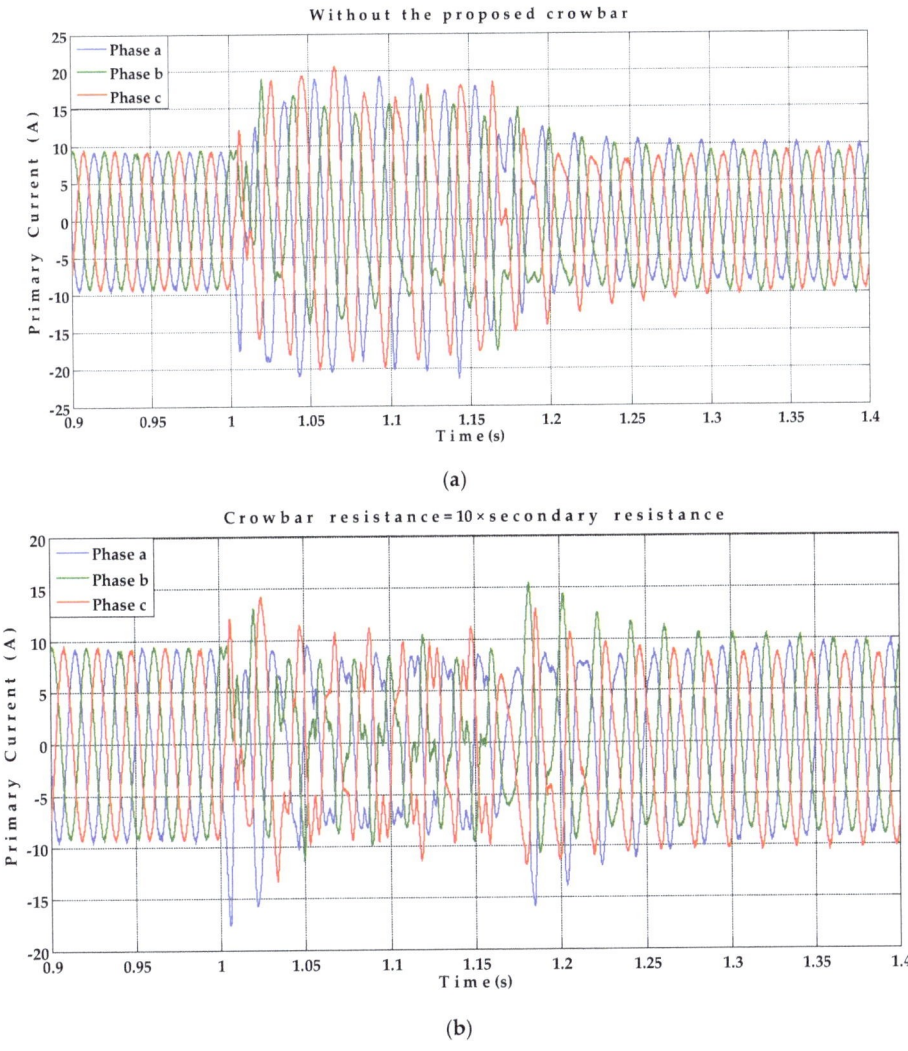

Figure 26. Primary current of BDFRG without and with the proposed crowbar under line to line fault occurrence: (**a**) Without the proposed crowbar; (**b**) With the proposed crowbar.

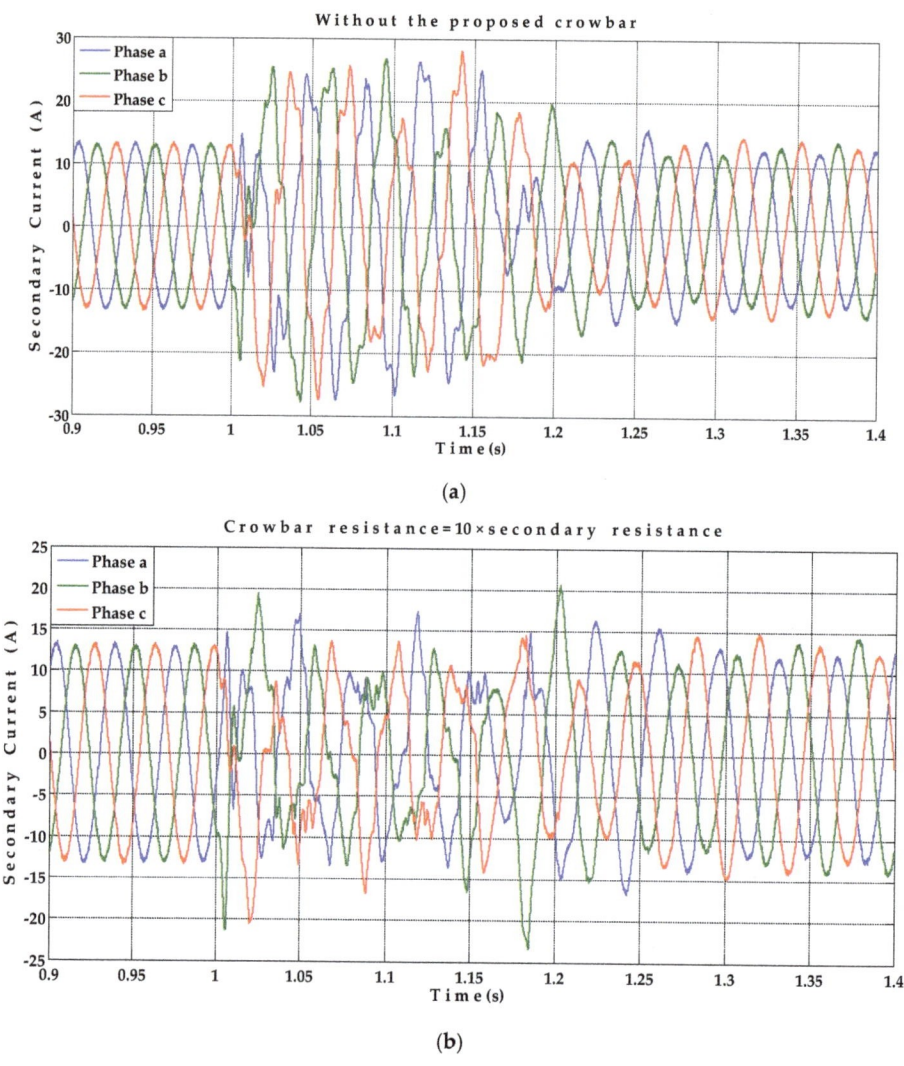

Figure 27. Secondary current of BDFRG without and with the proposed crowbar under line-to-line fault occurrence: (**a**) Without the proposed crowbar; (**b**) With the proposed crowbar.

The dc link voltage of the BDFRG (with and without using the proposed crowbar) is shown in Figure 28, which has a reference value equal to 710 V. Under the fault occurrence, in the case of "without using the proposed crowbar", the dc link voltage was decreased to about 584 V. In the case of using the proposed crowbar, the dc link voltage, during the fault, at first decreased to 690.5 V and then increased to about 777 V, then the dc link voltage improved and returned to its pre-fault value.

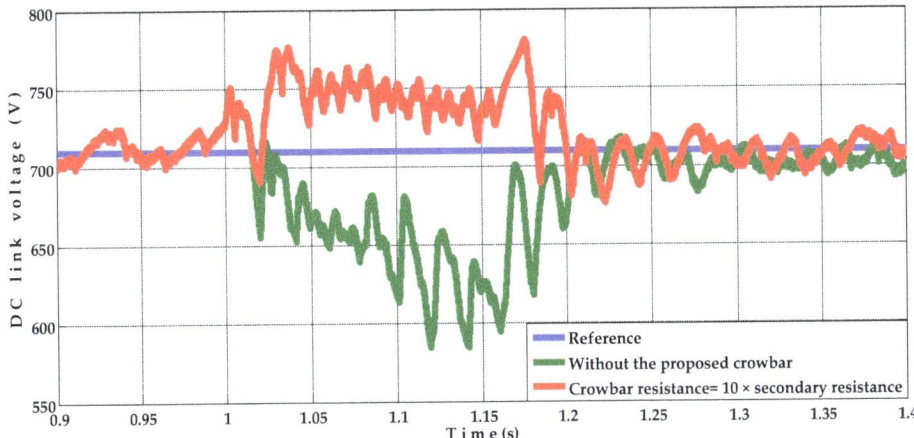

Figure 28. DC link voltage of BDFRG with and without the proposed crowbar under line-to-line fault occurrence.

5.2.3. Double Line to Ground Fault (The Fault Is Applied at Phases a and b)

The per unit rms terminal voltage (Va) of the BDFRG, as shown in Figure 29, at the instant of fault occurrence, the terminal voltage dropped to zero p.u. for 150 ms due to the occurrence of the studied double line to ground fault; then, after fault clearance, the terminal voltage returns to its original value (1 p.u.).

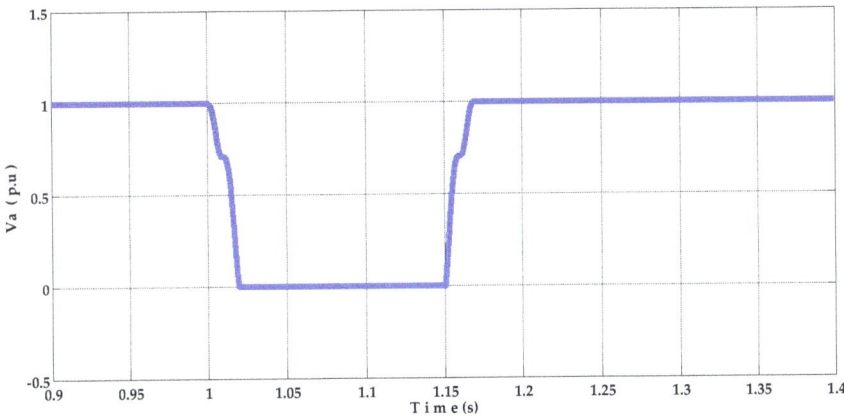

Figure 29. Per unit rms terminal voltage of the studied wind farm main coupling point under double line to ground fault occurrence.

The active power of the BDFRG (with and without using the proposed crowbar) is shown in Figure 30. As obvious in the case of "without using the proposed crowbar", during the fault, the active power dropped to 1.373 kW. In the case of using the proposed crowbar, during the fault, the active power was effectively improved and quickly returned to its pre-fault value.

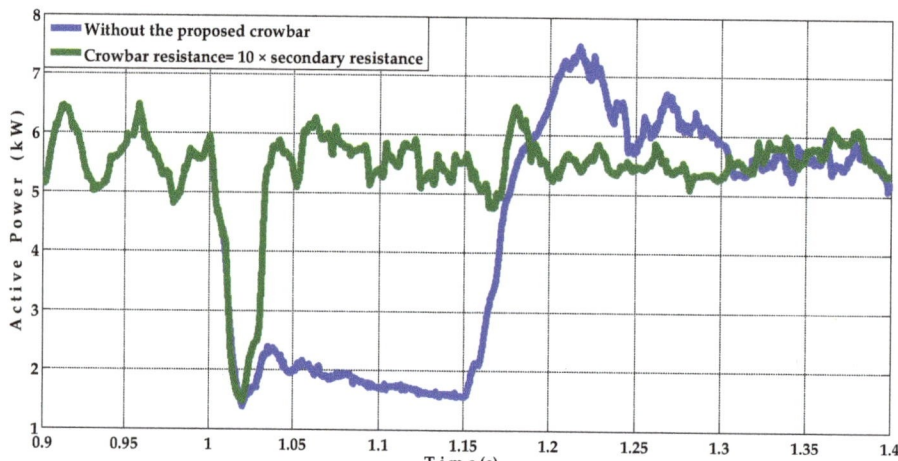

Figure 30. Active power of BDFRG with and without the proposed crowbar under double line to ground fault occurrence.

The reactive power of the BDFRG (with and without using the proposed crowbar) is shown in Figure 31. As shown, the reactive power was adjusted at zero value (unity power factor) before the fault occurrence. Following the clearance of the fault, in the case of "without using the proposed crowbar", the absorbed reactive power, by the BDFRG from the grid, reached about 5.15 kvar for a certain period. In the case of using the proposed crowbar, after fault clearance, the absorbed reactive power was reduced to 1.63 kvar only and quickly improved until reaching its pre-fault value.

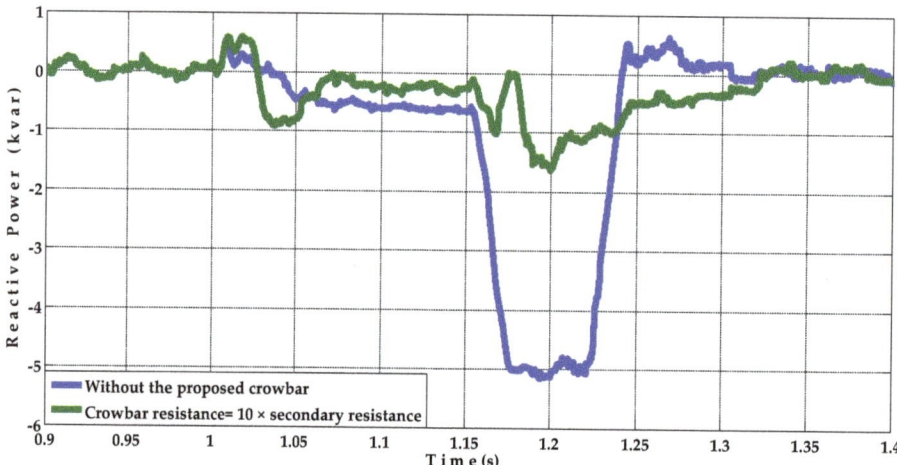

Figure 31. Reactive power of BDFRG with and without the proposed crowbar under double line to ground fault occurrence.

The rotor speed of the BDFRG (with and without using the proposed crowbar) is shown in Figure 32, which has a reference value equals to 1160 rpm. During the fault, in the case of "without using the proposed crowbar", the rotor rapidly accelerated; then, after fault clearance, the rotor speed reached about 1294 rpm, which led to a decrease in the power coefficient of the WT from 0.48 to less than 0.4605 as shown in Figure 33. While in the case of using the proposed crowbar, the rotor speed increased instantaneously to about 1206 rpm, then quickly improved.

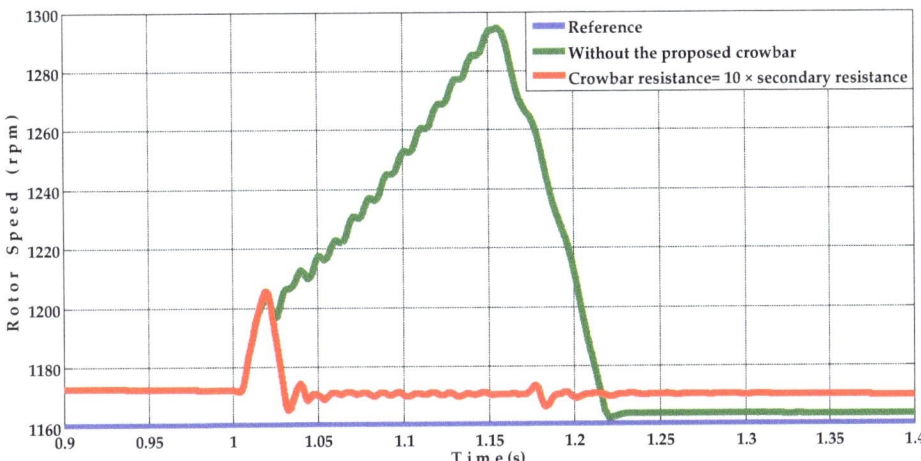

Figure 32. Rotor speed of BDFRG with and without the proposed crowbar under double line to ground fault occurrence.

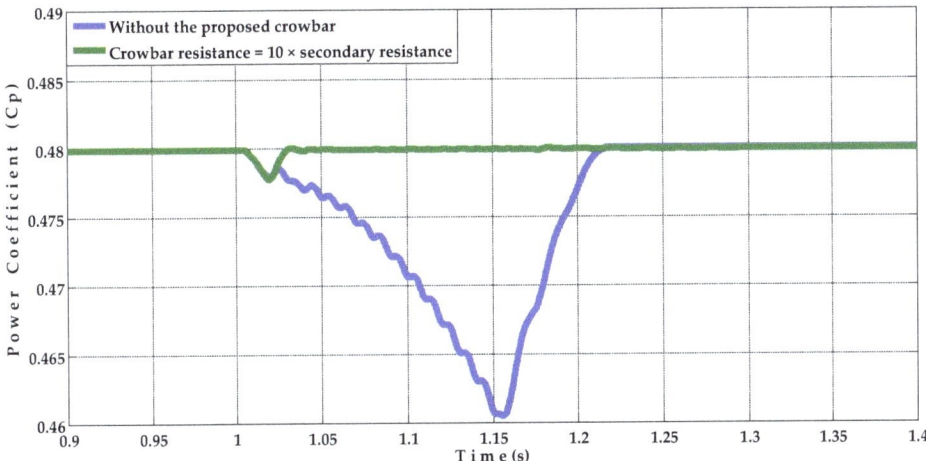

Figure 33. Power coefficient of wind turbine with and without the proposed crowbar under double line to ground fault occurrence.

The primary and secondary currents of the BDFRG (without and with using the proposed crowbar) are shown in Figures 34 and 35 in the same order. In the case of "without using the proposed crowbar", during the fault, both currents were increased; the primary current was increased to about 233% (21.69 A) of the pre-fault value (9.32 A) and the secondary current was increased to about 224% (29.29 A) of the pre-fault value (13.05 A), while in the case of using the proposed crowbar, as shown in Figures 34b and 35b, during the fault, both the primary and secondary currents were effectively improved.

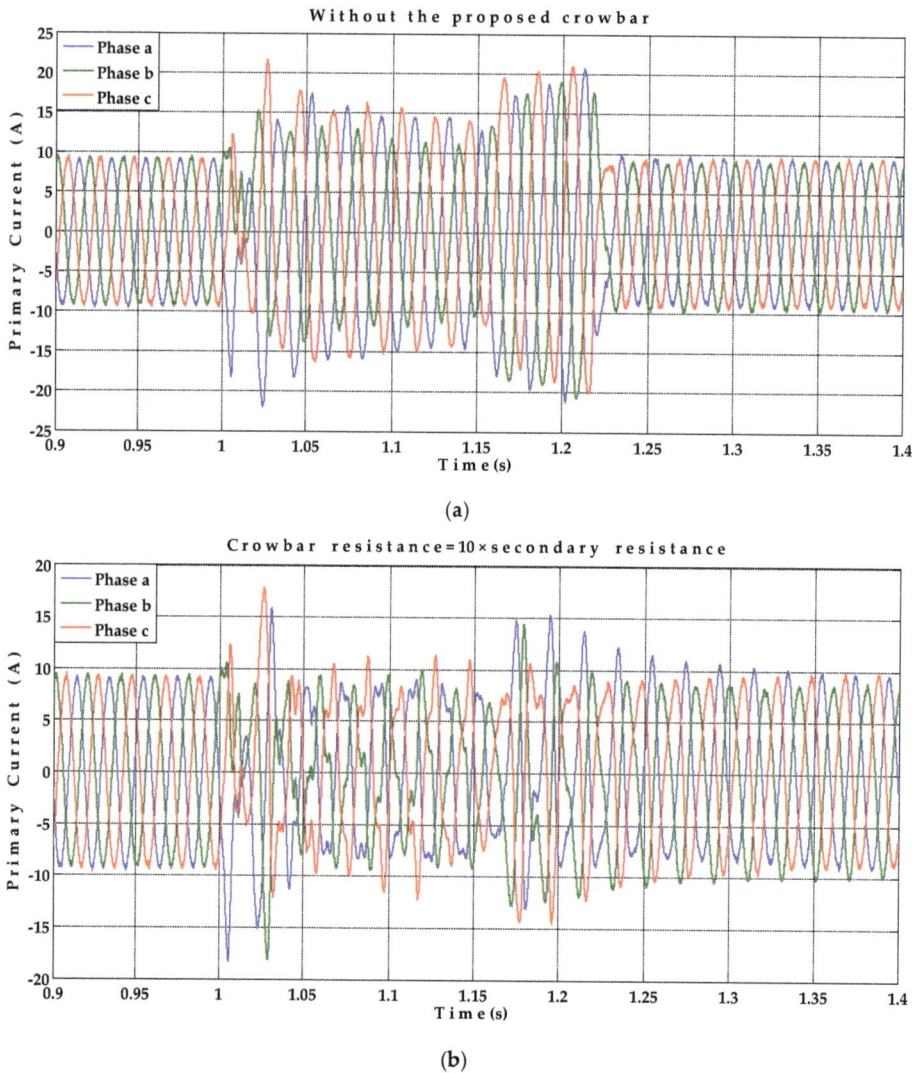

Figure 34. Primary current of BDFRG without and with the proposed crowbar under double line to ground fault occurrence: (**a**) Without the proposed crowbar; (**b**) With the proposed crowbar.

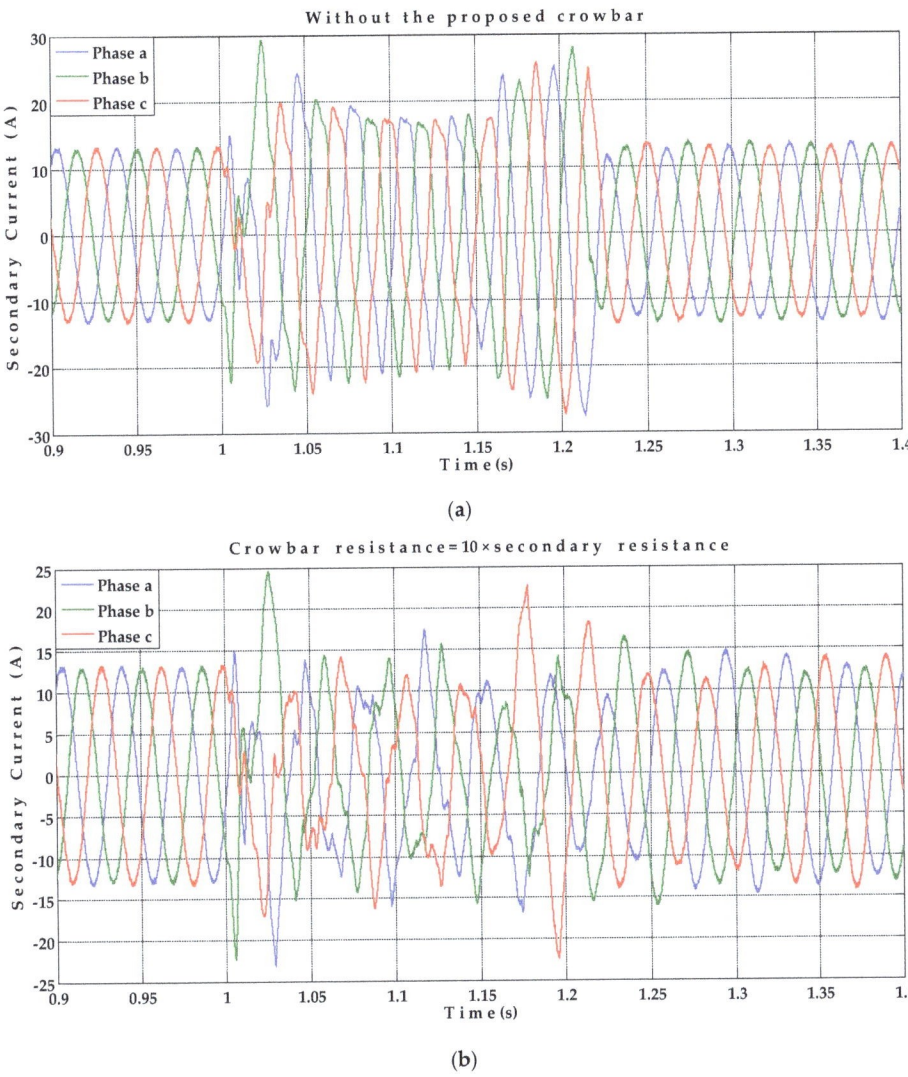

Figure 35. Secondary current of BDFRG without and with the proposed crowbar under double line to ground fault occurrence: (**a**) Without the proposed crowbar; (**b**) With the proposed crowbar.

The dc link voltage of the BDFRG (with and without using the proposed crowbar) is shown in Figure 36, which has a reference value equal to 710 V. Under the fault occurrence, in the case of "without using the proposed crowbar", the dc link voltage was decreased to about 595.5 V. In the case of using the proposed crowbar, the dc link voltage, during the fault, at first decreased to 667.2 V and then increased to about 766.4 V; then, the dc link voltage improved and returned to its pre-fault value.

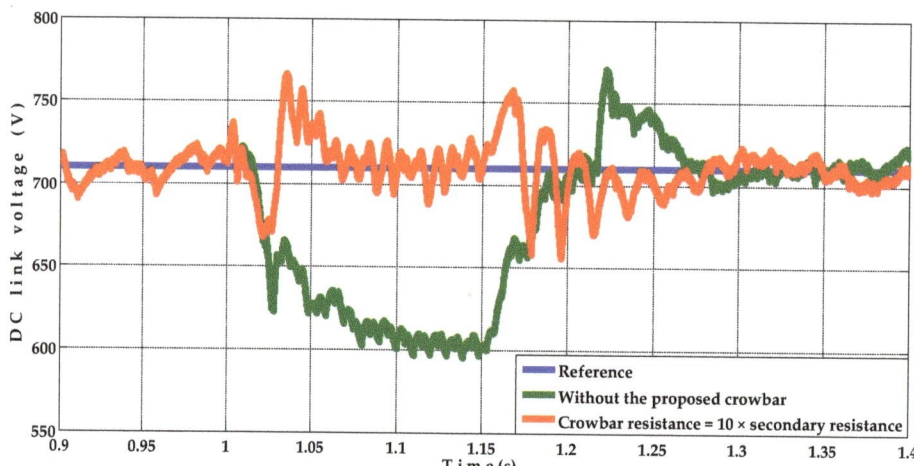

Figure 36. DC link voltage of BDFRG with and without the proposed crowbar under double line to ground fault occurrence.

6. Conclusions

To improve the capability of the BDFRG WTs to satisfy the grid code requirements concerning remaining the wind turbines connected to the grid under the occurrence of grid disturbances, many complex and expensive techniques were used such as using ANFIS control systems. This work proposes a new controllable crowbar as a new simple and economic solution to enhance the performance of the BDFRG WT under the occurrence of heavy faults. To examine the efficacy of the proposed controllable crowbar, the performance of the BDFRG WT under the occurrence of heavy different types of faults was studied twice, one without using the proposed controllable crowbar and the other with using it. To ensure that the proposed crowbar would be examined under the most extreme fault conditions, the location of the studied faults was chosen at the beginning of the transmission line next to the "wind farm" main point of common coupling. Not only that, but also to ensure accurate monitoring of the total actual performance of the BDFRG WT under the studied faults, all the protection system devices were deactivated. The terminal voltage, active power, reactive power, rotor speed, power coefficient, primary current of the BDFRG, secondary current of the BDFRG and DC link voltage were monitored and analyzed. Simulation results showed that in the case of fault occurrence without using the proposed controllable crowbar, all the monitored parameters were badly affected and the BDFRG WT would be rapidly disconnected from the network. On the other hand, simulation results showed that the proposed controllable crowbar effectively improved all the monitored parameters and enabled the studied BDFRG WT to remain in service under the studied faults.

Author Contributions: Conceptualization, M.R. and M.N.; methodology, M.R. and M.N.; software, M.R. and M.N.; validation, M.R., M.N. and A.E.-S.; formal analysis, M.R. and M.N.; investigation, M.R., M.N. and A.E.-S.; resources, M.R., A.E.-S. and M.N.; data curation, M.R. and M.N.; writing—original draft preparation, M.R. and M.N.; writing—review and editing, M.R., M.N. and B.H., A.E.-S.; visualization, M.R.; supervision, A.E.-S., M.R. and B.H.; project administration, A.E.-S.; funding acquisition, A.E.-S. All authors have read and agreed to the published version of the manuscript.

Funding: This research received no external funding.

Institutional Review Board Statement: Not applicable.

Informed Consent Statement: Not applicable.

Data Availability Statement: The data presented in this study are available on request from the corresponding author. The data are not publicly available due to their large size.

Conflicts of Interest: The authors declare no conflict of interest.

Nomenclature

List of Abbreviations

BDFRG	Brushless Doubly Fed Reluctance Generator
VSWT	Variable Speed Wind Turbine
WT	Wind Turbine
WECS	Wind Energy Conversion System
SCIG	Squirrel Cage Induction Generator
DFIG	Doubly Fed Induction Generator
BDFM	Brushless Doubly Fed Machine
BDFIM	Brushless Doubly Fed Induction Machine
BDFRM	Brushless Doubly Fed Reluctance Machine
BDFIG	Brushless Doubly Fed Induction Generator
MSC	Machine Side Converter
GSC	Grid Side Converter
MPPT	Maximum Power Point Tracking
SPWM	Sinusoidal Pulse Width Modulation
IFOC	Indirect Field Oriented Control
PLL	Phase-Locked Loop
RMS	Root Mean Square
ANFIS	Adaptive Neuro Fuzzy Inference System

List of Symbols

ω	speed of reference frame of power winding
ω_r	electrical speed of rotor
v_{dp}	direct voltage component for power winding
v_{qp}	quadrature voltage component for power winding
v_{dc}	direct voltage component for control winding
v_{qc}	quadrature voltage component for control winding
r_p	resistance of power winding
r_c	resistance of control winding
λ_{dp}	direct flux component for power winding
λ_{qp}	quadrature flux component for power winding
λ_{dc}	direct flux component for control winding
λ_{qc}	quadrature flux component for control winding
i_{dp}	direct current component for power winding
i_{qp}	quadrature current component for power winding
i_{dc}	direct current component for control winding
i_{qc}	quadrature current component for control winding
L_p	inductance of power winding
L_c	inductance of control winding
L_{pc}	mutual inductance between power and control winding
T_e	electrical torque produced from generator
p_r	number of poles for rotor
T_m	mechanical torque from turbine
n_g	turns ratio for gear box
w_{rm}	mechanical speed of rotor
J_r	moment of inertia for wind turbine
J_g	moment of inertia for generator
ω_p	angular speed for power winding
ω_s	angular speed for control winding
θ_p	primary flux angle
θ_g	grid current angle

Appendix A

Table A1. BDFRG parameters.

Rated Line Voltage	Rated Frequency	r_p	r_c	L_p	L_c	L_{pc}	Rotor Inertia, J_g
380 V	50 Hz	3.781 Ω	2.441 Ω	0.41 H	0.316 H	0.3 H	0.2 kg.m²

Table A2. Wind turbine parameters.

Rated Power	Turbine Radius, R	Wind Speed Range	Turbine Inertia, J_r	Gearbox Ratio, n_g
6 kW	4 m	2–12 m/s	1.5 kg. m²	7.5

References

1. Global Wind Energy Council. Global Wind Report 2022. Available online: https://gwec.net/ (accessed on 9 May 2022).
2. Wu, B.; Lang, Y.; Zargari, N.; Kouro, S. Fundamentals of Wind Evergy Conversion System Control. In *Power Conversion and Control of Wind Energy Systems*; Wiley-IEEE Press: Piscataway, NJ, USA, 2011; pp. 25–47. [CrossRef]
3. Sugirtha, M.G.; Latha, P. Analysis of power quality problems in grid connected wind power plant. In Proceedings of the 2011 International Conference on Recent Advancements in Electrical, Electronics and Control Engineering, Sivakasi, India, 15–17 December 2011; pp. 19–24. [CrossRef]
4. Mousa, M.G.; Allam, S.M.; Rashad, E.M. Maximum power tracking of a grid-connected wind-driven brushless doubly-fed reluctance generator using scalar control. In Proceedings of the 8th IEEE GCC Conference and Exhibition, Muscat, Oman, 1–4 February 2015; pp. 1–6. [CrossRef]
5. Allam, S.M.; Azmy, A.M.; El-Khazendar, M.A.; Mohamadein, A.L. Dynamic analysis of a BDFIM with a simple-proposed modification in the cage-rotor. In Proceedings of the 13th International Middle-East Power Systems Conference (MEPCON'2009), Cairo, Egypt, 13–15 December 2009; pp. 356–360.
6. Taluo, T.; Ristić, L.; Jovanović, M. Dynamic Modeling and Control of BDFRG under Unbalanced Grid Conditions. *Energies* 2021, 14, 4297. [CrossRef]
7. Moazen, M.; Kazemzadeh, R.; Azizian, M.-R. Mathematical Proof of BDFRG Model under Unbalanced Grid Voltage Condition. *Sustain. Energy Grids Netw.* 2020, 21, 100327. [CrossRef]
8. Kalaivani, M.; Kalappan, K.B. Design and Modeling of Brushless Doubly-Fed Reluctance Generator for Wind Mills. *Int. J. Recent Technol. Eng. (IJRTE)* 2019, 8, 74–80. [CrossRef]
9. Ademi, S.; Jovanovic, M.G.; Hasan, M. Control of Brushless Doubly-Fed Reluctance Generators for Wind Energy Conversion Systems. *IEEE Trans. Energy Convers.* 2015, 30, 596–604. [CrossRef]
10. Jin, S.; Shi, L.; Zhu, L.; Cao, W.; Dong, T.; Zhang, F. Dual Two-Level Converters Based on Direct Power Control for an Open-Winding Brushless Doubly-Fed Reluctance Generator. *IEEE Trans. Ind. Appl.* 2017, 53, 3898–3906. [CrossRef]
11. Moazen, M.; Kazemzadeh, R.; Azizian, M.R. Power control of BDFRG variable-speed wind turbine system covering all wind velocity ranges. *Int. J. Renew. Energy Res.* 2016, 6, 477–486. [CrossRef]
12. Kumar, M.; Das, S.; Kiran, K. Sensorless Speed Estimation of Brushless Doubly-Fed Reluctance Generator Using Active Power Based MRAS. *IEEE Trans. Power Electron.* 2019, 34, 7878–7886. [CrossRef]
13. Kumar, M.; Das, S. Model reference adaptive system based sensorless speed estimation of brushless doubly-fed reluctance generator for wind power application. *IET Power Electron.* 2018, 11, 2355–2366. [CrossRef]
14. Moazen, M.; Kazemzadeh, R.; Azizian, M.R. Model-based predictive direct power control of brushless doubly fed reluctance generator for wind power applications. *Alex. Eng. J.* 2016, 55, 2497–2507. [CrossRef]
15. Rihan, M.; Nasrallah, M.; Hasanin, B. Performance analysis of grid-integrated brushless doubly fed reluctance generator-based wind turbine: Modelling, control and simulation. *SN Appl. Sci.* 2020, 2, 114. [CrossRef]
16. Tsili, M.; Papathanassiou, S. A review of grid code technical requirements for wind farms. *IET Renew. Power Gener.* 2009, 3, 308–332. [CrossRef]
17. Wu, Q.; Sun, Y. Grid Code Requirements for Wind Power Integration. In *Modeling and Modern Control of Wind Power*; Wiley-IEEE Press: Piscataway, NJ, USA, 2017; pp. 20–53. [CrossRef]
18. Hu, Y.-L.; Wu, Y.-K.; Chen, C.-K.; Wang, C.-H.; Chen, W.-T.; Choc, L.-I. A Review of the Low-Voltage Ride-Through Capability of Wind Power Generators. In Proceedings of the 4th International Conference on Power and Energy Systems Engineering, CPESE 2017, Berlin, Germany, 25–29 September 2017. [CrossRef]
19. Wind Farm Grid Connection Code in Addition to the Egyptian Transmission Grid Code (ETGC). Available online: https://www.eehc.gov.eg/eehcportalnew/NewEnergyPDF/Egypt_gridcode_for_wind_farm_connection.pdf (accessed on 9 May 2022).
20. Duong, M.Q.; Leva, S.; Mussetta, M.; Le, K.H. A Comparative Study on Controllers for Improving Transient Stability of DFIG Wind Turbines During Large Disturbances. *Energies* 2018, 11, 480. [CrossRef]
21. Long, T.; Shao, S.; Malliband, P.; Abdi, E.; McMahon, R.A. Crowbarless Fault Ride-Through of the Brushless Doubly Fed Induction Generator in a Wind Turbine Under Symmetrical Voltage Dips. *IEEE Trans. Ind. Electron.* 2013, 60, 2833–2841. [CrossRef]

22. Hansen, A.D.; Michalke, G. Fault ride-through capability of DFIG wind turbines. *Renew. Energy* **2007**, *32*, 1594–1610. [CrossRef]
23. Abad, G.; Rodriguez, M.; Poza, J.; Canales, J.M. Direct torque control for doubly fed induction machine-based wind turbines under voltage dips and without crowbar protection. *IEEE Trans. Energy Convers.* **2010**, *25*, 586–588. [CrossRef]
24. Xiang, D.; Ran, L.; Tavner, P.J.; Yang, S. Control of a doubly fed induction generator in a wind turbine during grid fault ride-through. *IEEE Trans. Energy Convers.* **2006**, *21*, 652–662. [CrossRef]
25. Betz, R.E.; Jovanovic, M.G. *Introduction to Brushless Doubly Fed Reluctance Machines. The Basic Equations*; Alborg University: Alborg, Denmark, 1998. [CrossRef]
26. Jovanovi, M.; Chaal, H. High-Performance Control of Doubly-Fed Reluctance Machines. *Electronics* **2010**, *14*, 72–77.
27. Ademi, S.; Jovanovid, M.; Obichere, J.K. Comparative analysis of control strategies for large doubly-fed reluctance wind generators. In Proceedings of the World Congress on Engineering and Computer Science WCECS 2014, San Francisco, CA, USA, 22–24 October 2014.
28. Rihan, M. Comparison among Different Crowbar Protection Techniques for Modern Wind Farms under Short Circuit Occurrence. In Proceedings of the 19th International Middle East Power Systems Conference (MEPCON'19), Cairo, Egypt, 19–21 December 2017; pp. 1113–1121. [CrossRef]
29. Noureldeen, O.; Hamdan, I. An efficient ANFIS crowbar protection for dfig wind turbines during faults. In Proceedings of the 19th International Middle East Power Systems Conference (MEPCON'19), Cairo, Egypt, 19–21 December 2017; pp. 263–269. [CrossRef]
30. Noureldeen, O. Behavior of DFIG Wind Turbines with Crowbar Protection under Short Circuit. *Int. J. Electr. Comput. Sci. IJECS-IJENS* **2012**, *12*, 32–37.

Article

Single-Phase Universal Power Compensator with an Equal VAR Sharing Approach

Nishant Patnaik [1], Richa Pandey [2], Raavi Satish [3], Balamurali Surakasi [3], Almoataz Y. Abdelaziz [4] and Adel El-Shahat [5,*]

1. Department of Electrical and Electronics Engineering, Chaitanya Bharathi Institute of Technology, Hyderabad 500075, India; nishanthpatnaik_eee@cbit.ac.in
2. Department of Electrical and Electronics Engineering, MVGR College of Engineering, Vizianagaram 530048, India; richap68@gmail.com
3. Department of Electrical and Electronics Engineering, Anil Neerukonda Institute of Technology and Sciences, Visakhapatnam 531162, India; satish.eee@anits.edu.in (R.S.); balamurali.eee@anits.edu.in (B.S.)
4. Faculty of Engineering and Technology, Future University in Egypt, Cairo 11835, Egypt; almoataz.abdelaziz@fue.edu.eg
5. Energy Technology Program, School of Engineering, Purdue University, West Lafayette, IN 47906, USA
* Correspondence: asayedah@purdue.edu

Abstract: In this manuscript, we propose a single-phase UPC (universal power compensator) system to extensively tackle power quality issues (voltage and current) with an equal VAR (volt-ampere reactive) sharing approach between the series and shunt APF (active power filter) of a UPC system. The equal VAR sharing feature facilitates the series and shunt APF inverters to be of an equal rating. An SRF (synchronous reference frame)-based direct PA (power angle) calculation technique is implemented to realize equal VAR sharing between the APFs of the UPC. This PA estimation utilizes d and q axis current parameters derived for the reference signal generation of the shunt APF. An SRF-based method is highly useful for power estimations in distorted supply voltage conditions compared with other conventional methods, i.e., the PQ method. It comprises a reduced complexity and estimations with an easiness to retain two APF inverters of equal rating. A rigorous simulation analysis is performed with MATLAB/SIMULINK and a real-time digital simulator (OPAL-RT) for addressing different power quality-disturbing elements such as current harmonics, voltage harmonics, voltage sag/swell and load VAR demand with the proposed method.

Keywords: APF; power quality; SRF; UPC; VAR

1. Introduction

In the present age of computerization and digitalization, complex devices with fast processing controllers draw current shapes that are unwanted in the supply system. This significantly degrades the quality of the power (i.e., voltage and current) supplied or available for other loads [1–5]. A specific conditioning equipment known as a unified power quality conditioner (UPQC) can be employed for such types of loads to resolve both voltage- and current-related issues, thus assisting with the efficient operation of the load and preventing the supply lines from being infected with the non-linearities and reactive power demand of the load [6–21].

Various UPQC control methods considering different compensating scenarios have been considered by the research community. Along with regular voltage and current compensation, reactive power compensation by a device has also proven to be a major utilization [11]. The ability of reactive power compensation has been explored further in order to be shared among the shunt and series APFs of the device by using the power angle control (PAC) concept [11–14,18,20]. It has been effectively implemented in a three-phase three-wire system [11,12] and a three-phase four-wire system [14,20]. The PAC concept has also been implemented in a single-phase system-based UPQC, as discussed in [18].

The estimation of the PA (power angle), as discussed in the literature, is based on the PQ concept, primarily for three-phase systems utilizing power components to calculate the PA. It has already been well-established that, under non-sinusoidal supply voltage conditions, PQ concept-based harmonic compensation exhibits a poor performance compared with the SRF concept [22]. Thus, under similar conditions, a PA estimation with a PQ concept also followed a similar trail, as discussed in the manuscript. It also necessitated the regular computation of VAR sharing by each APF for the computation of the PA [11]. This regular VAR computation can be avoided by using two APFs with an equal rating, thus sharing equal reactive power.

In this paper, we present a comprehensive compensating device, termed a UPC (universal power compensator), for a single-phase system with equal APF ratings whose control action exhibits a few important features. One is the realization of equal reactive power sharing between both APFs, thus avoiding a regular VAR computation. Secondly, the equal VAR sharing is comprehended by the PA estimation without power estimations, directly utilizing the SRF current parameters (i_d-i_q), which are estimated for current harmonic compensation. Thirdly, as only the d-q current parameters are used for the PA estimation, it remains unfazed due to the non-sinusoidal supply voltage condition. To demonstrate its effectiveness, a comparative analysis is presented in reference to PQ concept-based compensation and the PA estimation under a non-sinusoidal supply voltage condition.

A single-phase UPC with a shunt and series APF combined through a DC-link capacitor is shown in Figure 1. The specific load is considered to be an arrangement of the linear and non-linear load. The non-linear load comprises a single-phase diode bridge rectifier with an RL load on the DC side whereas the linear load consists of only a series combination of the active and reactive load. With both loads in operation, the source current harmonics are not as high, but they have a considerable VAR demand that needs to be compensated for on an equal sharing basis by the APFs of the UPC. When the non-linear load acts alone, it contaminates the source current with the harmonics to a greater extent, which is tackled by the shunt APF. Voltage sag, voltage swell and voltage harmonics, or the supply side disturbances, should be mitigated by the series APF.

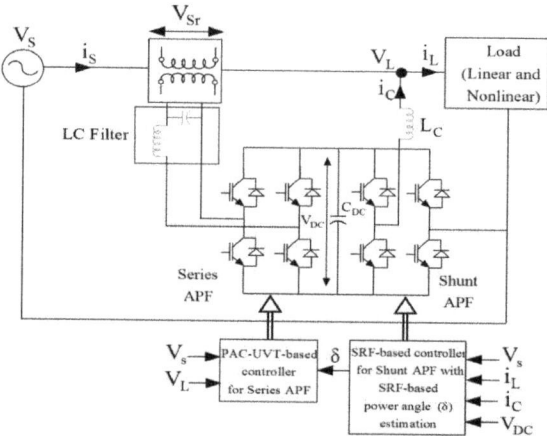

Figure 1. Single-phase UPC system with equal reactive power sharing approach.

2. Shunt APF Reference Signal Generation with SRF-Based Approach

By an orthodox approach of VAR sharing between the shunt and series APFs as presented in [4,5], the series APF part of handling VAR arose after the load VAR surpassed a specific maximum.

The shunt APF of the UPC was used for serving the following two purposes in our system:

1. Providing source current harmonic compensation.
2. Providing 50% of load reactive power compensation.

The SRF control algorithm is a popular control approach for custom power devices as it involves the direct controlling of the d (active) and q (reactive) elements of the current drawn by the load. As shown in Figure 2, initially, the single-phase load current was treated as a load component $i_{L\alpha}$ in stationary reference frame; the other component $i_{L\beta}$ was derived by orthogonally shifting $i_{L\alpha}$. The transformation angle ωt was generated from the modified PLL (phase-locked loop) due to the source voltage with the harmonics for a proper synchronization [14]. With the following transformations, the direct and quadrature components of the current in a rotating reference frame were obtained from $i_{L\alpha}$ and $i_{L\beta}$.

$$\begin{bmatrix} i_{Ld} \\ i_{Lq} \end{bmatrix} = \begin{bmatrix} \cos \omega t & \sin \omega t \\ -\sin \omega t & \cos \omega t \end{bmatrix} \begin{bmatrix} i_{L\alpha} \\ i_{L\beta} \end{bmatrix} \quad (1)$$

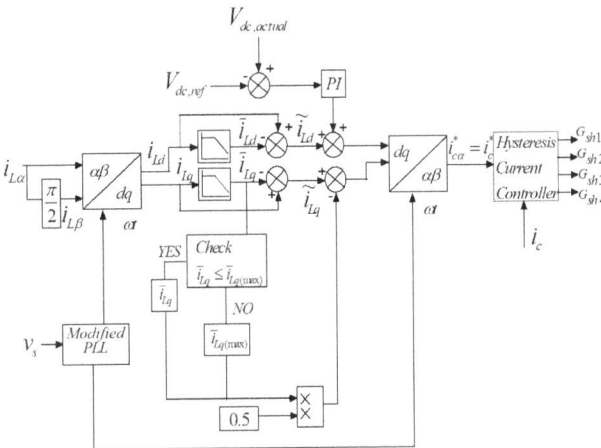

Figure 2. SRF method-based shunt APF.

The direct component of the current, i.e., i_{Ld}, was considered to be an active component of the current whereas the quadrature component of the current i_{Lq} was considered to be a reactive element of the load current.

These two components involved fundamental and harmonic terms and could be separated by a different HPF (high-pass filter) derived from an LPF (low-pass filter). The loss element of the UPC, acquired from a PI compensator by comparing the DC-link reference voltage with the measured DC-link voltage, was combined with the harmonic content of the active current. The VAR harmonic current element was combined with 50% of the fundamental reactive current to allow half of the load VAR to be remunerated by the shunt APF. The compensating d and q axis currents thus obtained were transformed to a reference shunt APF-injecting current. This current was compared with the measured shunt APF current and fed to a hysteresis controller to produce a gating signal for the shunt APF.

3. PA Calculation for Equal Reactive Power Sharing

Considering the non-ideal conditions of the supply voltage and the load current, the d and q axis voltage and current parameters, obtained from the SRF control algorithm of the shunt APF, were utilized to estimate the power demand by load as:

$$\text{Load active power (fundamental)}, P_L = \bar{v}_{sd} \cdot \bar{i}_{Ld} \quad (2)$$

$$\text{Load reactive power (fundamental)}, Q_L = \bar{v}_{sd} \cdot \bar{i}_{Lq} \quad (3)$$

The source side power could be calculated as:

$$\text{Source active power, } P_S = \bar{v}_{sd} \cdot (\bar{i}_{Ld} + \bar{i}_0) \quad (4)$$

$$\text{Source reactive power, } Q_S = 0 \quad (5)$$

where $\bar{i}_0$, the active component of the current drawn from the source, denotes the loss component of the UPC.

The VAR rating of the series APF could be articulated as [4]:

$$Q_{sr} = |I_s| \cdot |V_L^*| \cdot \sin\delta \quad (6)$$

$$\sin\delta = \frac{Q_{sr}}{|I_s| \cdot |V_L^*|} = \frac{Q_{sr}}{P_S} \quad (7)$$

$$\delta = \sin^{-1}\left[\frac{Q_{sr}}{P_s}\right] \quad (8)$$

For equivalent VAR sharing among the shunt and series APF we obtained:

$$Q_{sr} = 0.5 Q_L \quad (9)$$

Therefore, from Equations (3), (4), (8) and (9), we obtained:

$$\delta = \sin^{-1}\left[\frac{i_{Lq}}{2 \cdot (\bar{i}_{Ld} + \bar{i}_0)}\right] \quad (10)$$

Thus, as is clear from the above equation, for half of the load VAR sharing by the series APF, the PA could be directly derived from the load current parameters. As the estimation of the PA was independent of any voltage component, in reference to the PA estimation presented in [11,12] (where the PA estimation was based on the PQ theory), the number of variables involved were reduced in the PA estimation. Therefore, the PA estimation was much simpler to implement. In [12], a PA estimation was carried out where the supply voltage was considered to be completely sinusoidal. With a non-sinusoidal supply voltage and the methods proposed in [11,12], an additional disturbance needs to be accounted for, as discussed in Section 4. A comparative analysis is presented in Section 5 to understand the efficacy of the SRF-based PA estimation over the others.

4. Comparison of SRF-Based Control with a Conventional PQ (Active-Reactive Power)-Based Technique

The PQ (active-reactive power) method of control for compensation by an active power filter (APF) is one of the most popular control methods. It is based on separating the average and harmonic power components. However, the SRF method is based on separating the average and harmonic components of the current. Under ideal sinusoidal supply conditions, both the SRF method and the conventional PQ method perform similarly. However, under non-sinusoidal supply conditions, the compensating power consists of additional disturbing factors in the PQ method (id-iq). Thus, under non-ideal supply voltage conditions, the SRF method is always a preferable option compared with the PQ method [5,22].

Under a non-sinusoidal supply, the compensating power by the APF for the PQ method is given by:

$$\begin{bmatrix} P_{C(PQ)} \\ Q_{C(PQ)} \end{bmatrix} = -\bar{v}_{sd}\begin{bmatrix} \tilde{i}_{Ld} \\ -\tilde{i}_{Lq} \end{bmatrix} - \tilde{v}_{sd}\left(\begin{bmatrix} \tilde{i}_{Ld} \\ -\tilde{i}_{Lq} \end{bmatrix} + \begin{bmatrix} \bar{i}_{Ld} \\ -\bar{i}_{Lq} \end{bmatrix}\right) \quad (11)$$

Under a non-sinusoidal supply, the compensating power by the APF for the SRF method is given by:

$$\begin{bmatrix} P_{C(SRF)} \\ Q_{C(SRF)} \end{bmatrix} = -(\bar{v}_{sd} + \tilde{v}_{sd}) \begin{bmatrix} \tilde{i}_{Ld} \\ -\tilde{i}_{Lq} \end{bmatrix} \quad (12)$$

Thus, the difference between Equations (11) and (12) accounts for the additional disturbance in the PQ method. Another important degrading factor in the PQ method may arise if similar harmonic components are present in the voltage and current, resulting in a more fundamental power component.

In Section 6.4, the performance difference is presented between the PQ method and the SRF method under non-sinusoidal voltage disturbances.

5. PA-UVT Controller for Series APF

As loads operate at a rated voltage, the reference voltage could be fixed and a unit vector template (UVT) generation method could be employed for the series APF, as illustrated in Figure 3. For incorporating the PA shift in the load voltage in reference to the source voltage, the transformation angle ωt was added with the PA δ. The transformation angle ωt was generated from the modified PLL, as stated in Section 2. The reference load voltage signal generated is depicted below:

$$V_L^* = V_m \sin(\omega t + \delta) \quad (13)$$

where V_m is the maximum fixed reference rated voltage. The desired load voltage (V_L^*) was equated with the measured load voltage (V_L) to produce the control signals for the series APF.

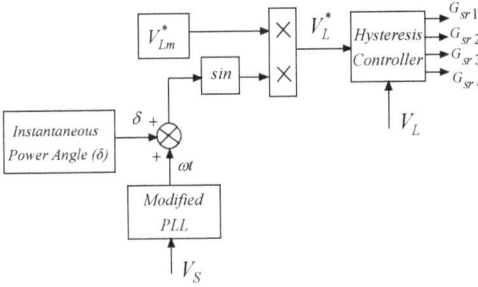

Figure 3. PAC-SRF-based series APF.

6. Simulation Results

This new sharing algorithm-based UPC system was implemented in SIMULINK for an analysis considering the various operational conditions of the source and load. The three different types of disturbances in the source voltage were sag, swell and harmonics. Correspondingly, on the load side, the performance of the UPC was analyzed with three various load situations; i.e., with a non-linear load only, a linear load only and a composite type (both linear and non-linear). The parameters considered for the simulation analysis are depicted in Table 1. With source voltage disturbances being common phenomena, the simulation results were analyzed under different categories of loading conditions. The simulation time was separated into various time divisions as per the source voltage disturbance. From 0 to 0.3 s, the source voltage was in a steady state rated condition; from 0.3 to 0.6 s, a voltage sag of 20% was introduced; from 0.6 to 0.9 s, a voltage swell of 20% was incorporated along with the rated source voltage; and from 0.9 to 1.2 s, the 3rd, 5th and 7th harmonic voltage component (10% each) was injected into the steady state voltage.

Table 1. Parameters for simulation and real-time analysis.

Supply system	Source impedance	R = 0.06 Ω, L = 0.05 mH
	Source voltage	230 V (RMS), 50 Hz
Shunt APF	Coupling inductor	2.0 mH
	PI parameters	K_P = 0.2, K_I = 10.34
Series APF	Filter components	L = 1.5 mH, C = 65 µF
	Transformer	1:1, 3 kVA
DC-link	Capacitor	1600 µF
	Desired voltage	350 V
Load system	Linear load	P_{max} = 5 kW, Q_{max} = 5 kVAR
	Non-linear load	One-phase diode bridge rectifier with R = 8 Ω, L = 10.5 mH

6.1. Simulation Results with a Composite Load

In the presence of both a linear and non-linear load, the THD content of the source current without compensation was less than that obtained with the non-linear load alone, but remained on the higher side. With a constant maximum linear load of 5 kW and 5 kVAr, the series and shunt APFs took part equally in reactive power compensation, along with their respective voltage disturbance and current harmonic compensation. Figures 4 and 5 depict the different voltage and current waveforms under this load condition, respectively. Figure 4a illustrates the voltage waveforms for the complete duration of 1.2 s. Figure 4b–d depicts enlarged fragments of the voltage waveform observing the sag, swell and harmonic condition, respectively. It was observed that, under all scenarios, the load voltage was maintained as the same as the source voltage before any disturbance; i.e., before 0.3 s. Likewise, Figure 5b–d depicts enlarged fragments observing the sag, swell and harmonic condition, respectively, of the total current waveform shown in Figure 5a. The DC-link voltage profile was also maintained at its set reference value, as can be observed in Figure 6. Although a DC-link voltage ripple was observable, the UPC performance was not affected and was satisfactory with the considered operating conditions.

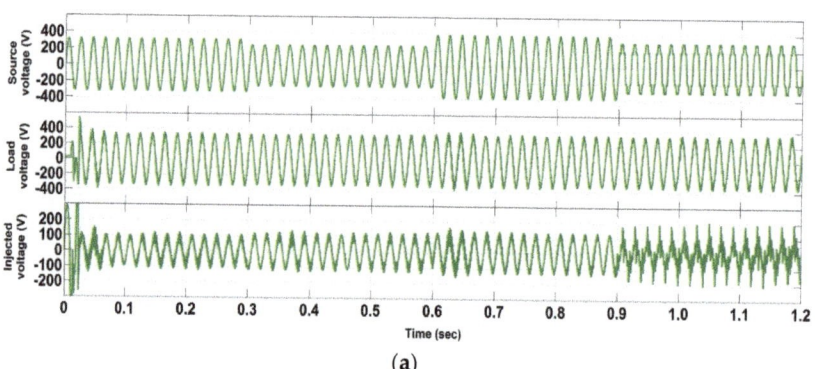

(a)

Figure 4. Cont.

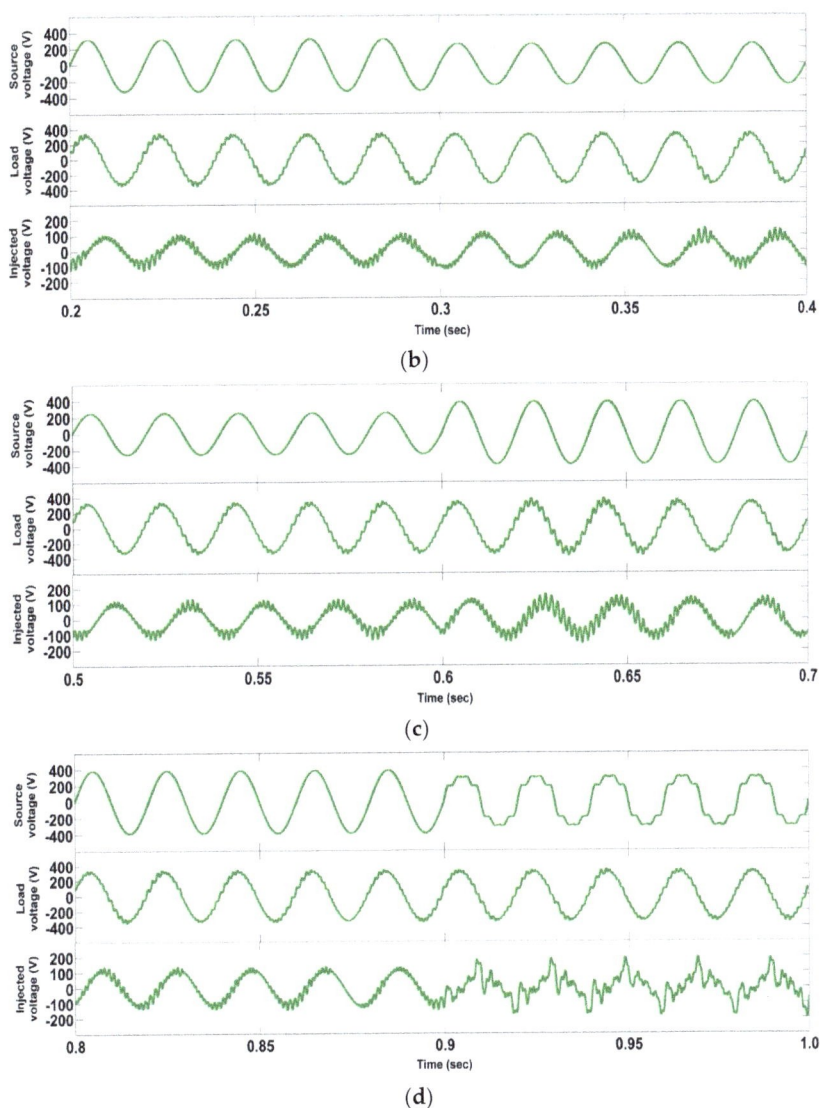

Figure 4. Simulation results with composite load (linear and non-linear): source voltage, load voltage and series injected voltage (top to bottom) at (**a**) 0–1.2 s, (**b**) 0.2–0.4 s, (**c**) 0.5–0.7 s and (**d**) 0.8–1.0 s.

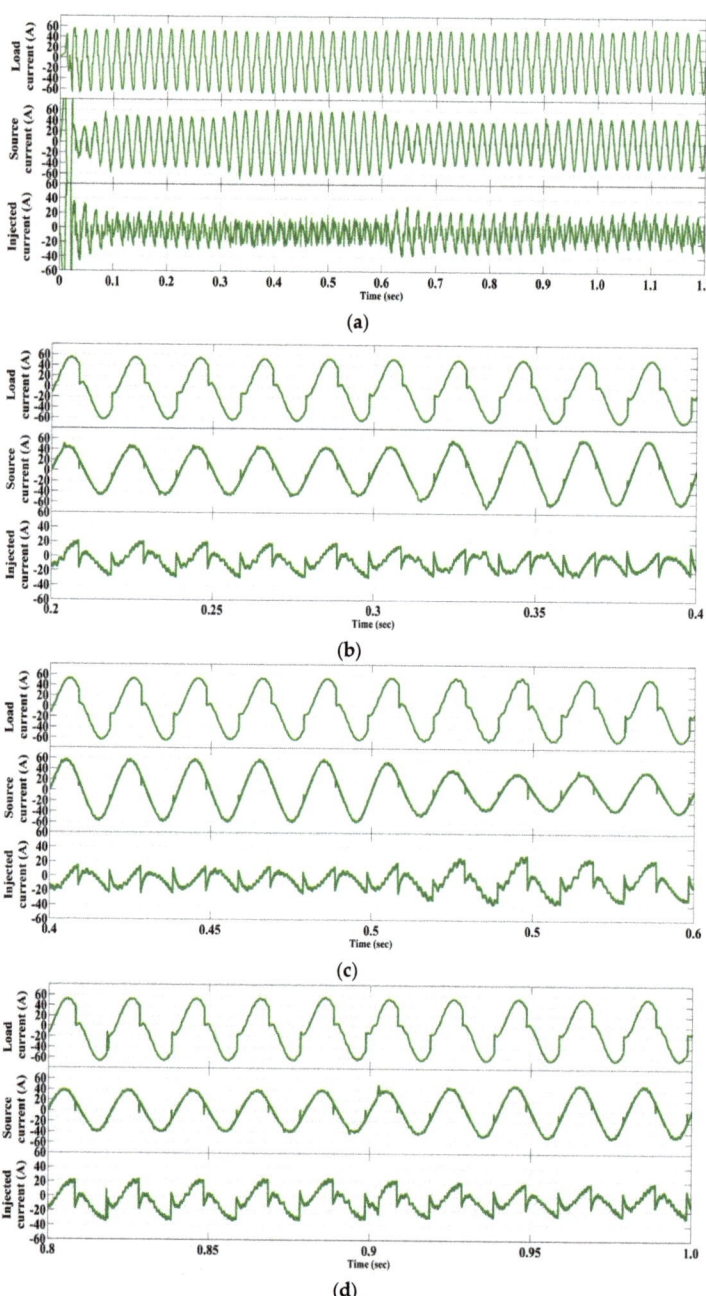

Figure 5. Simulation results with composite load (linear and non-linear): source current before compensation, source current after compensation and injection current from shunt APF (top to bottom) at (**a**) 0–1.2 s, (**b**) 0.2–0.4 s, (**c**) 0.5–0.7 s and (**d**) 0.8–1.0 s.

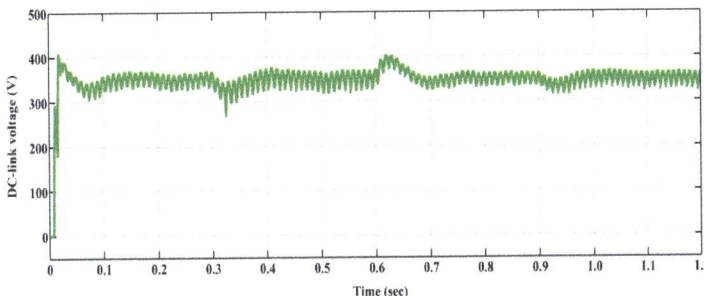

Figure 6. Simulation result with composite load (linear and non-linear): DC-link voltage.

6.2. Simulation Results with Linear Load

In the absence of any non-linear load, the source current without compensation was free from harmonics and did not require any harmonic compensation from the shunt APF. Figure 7 illustrates the response of the UPC system with the proposed controller for a constant linear load of 5 kW and 5 kVAR. Under varying source voltage disturbances, the series injected voltage made adjustments to maintain a steady state and rated load voltage profile. The compensating current from the shunt APF was only responsible for partly compensating the VAR demand whereas an equal amount of reactive power was compensated for by the series APF.

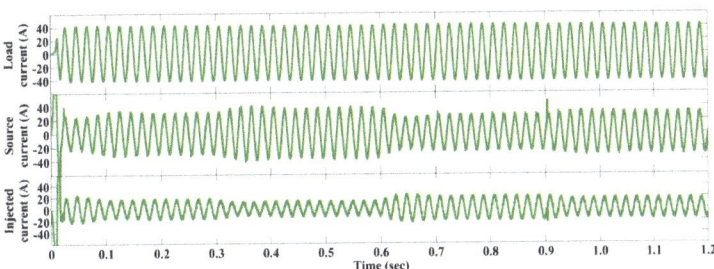

Figure 7. Simulation results with linear load only: source current before compensation, source current after compensation and injection current from shunt APF (top to bottom).

6.3. Simulation Results with a Non-Linear Load

With the inclusion of only a non-linear load, the UPC system operation was confined to the compensation of source voltage disturbances and current harmonic compensation. As is clear from Figure 8, the source current profile was free from harmonics, with the THD content being reduced to 3.4% from 23.6%. Figure 8a depicts the time duration of 1.2 s whereas Figure 8b is an enlarged fragment for the time duration of 0.3–0.6 s for a better illustration of the effect of the non-linear load on the current waveform. It was evident that the source current before compensation was highly contaminated with harmonics whereas after compensation it was closer to sinusoidal. The THD analysis of the source current is discussed in detail further on in this work. With a proper compensating series injected voltage by the series APF and a compensating current by the shunt APF, the load voltage profile was maintained at its rated condition, as illustrated in Figure 9.

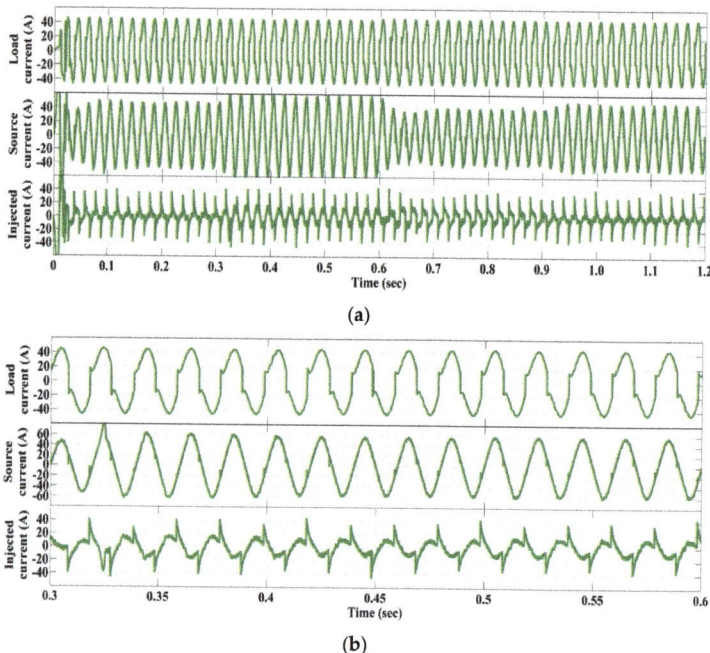

Figure 8. Simulation results with non-linear load only: source current before compensation, source current after compensation and injection current from shunt APF (top to bottom) at (**a**) 0–1.2 s and (**b**) 0.3–0.6 s.

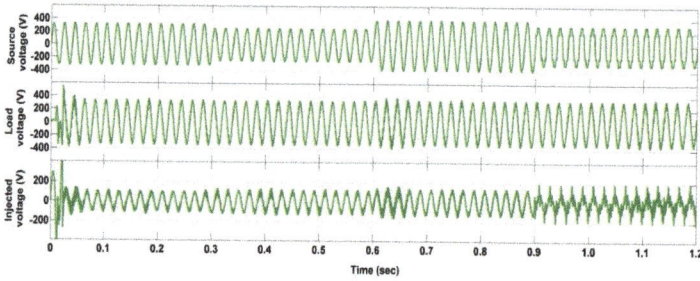

Figure 9. Simulation results with non-linear load only: source voltage, load voltage and series injected voltage (top to bottom).

6.4. Cumulative UPC Performance Parameters under Different Voltages and Loading Conditions

Table 2 illustrates the various performance parameters used to obtain a better understanding of the UPC performance analysis under different voltage and loading conditions. The performance parameters listed in the table are: I: the RMS value of the load voltage in volts (without a UPC); II: the RMS value of the load voltage (with a UPC); III: the THD% of the load voltage (without a UPC); IV: the THD% of the load voltage (with a UPC); V: the THD% of the source current (without a UPC); VI: the THD% of the source current (with a UPC); and VII: the reactive power share between the shunt APF (Q_{SH}) and the series APF (Q_{SR}) in VAR (volt-ampere reactive). As observed from the table, the load voltage RMS value was maintained close to the reference RMS value of 230 V under different voltage and loading conditions. The THD% of the load voltage with the harmonics under consideration was within the limits of 5% with an appropriate compensation from the UPC in all cases. As can be seen from the table, the THD% of the source current without the UPC, considering both a linear and non-linear load (composite load), was in the range of 9–10% whereas for a non-linear load, it was in the range of 16–20%. However, with the UPC, the THD% of the source current was brought back to within a 5% limit. Thereafter, reactive power compensation sharing by the UPC APFs were observed for the composite load and linear load cases. With the composite load, the total load reactive power demand was around 6 kVAR and equal sharing was observed by the shunt and series APF. With only a linear load, the load reactive power demand was 5 kVAR and, consequently, there was equal sharing between the shunt and series APF.

6.5. Comparative Performance Analysis under Non-Sinusoidal Voltage Conditions

A simulation case was considered again with a non-linear load, but with a non-sinusoidal supply voltage condition only. The focus was on the current compensation by the UPC under non-sinusoidal voltage conditions. Figure 10a,b indicates the source voltage with harmonics (10% of 3rd, 5th and 7th order harmonic components) and the load voltage with compensation from the UPC system. Figure 10c,d illustrates the FFT analysis of the source voltage and load voltage, respectively. The load voltage was reduced from around 15% to 3%.

The efficacy of the SRF method of compensation over the conventional PQ method under non-sinusoidal voltage conditions in terms of current compensation and equal reactive power sharing, as discussed in Section 4, is exhibited in a further analysis.

A. THD Analysis of Source Current between SRF and Conventional PQ Methods

Figure 11a shows the source current waveform obtained from the PQ method. The source current waveform after compensation from the SRF method is illustrated in Figure 11b. It was clearly observable that the source current waveform obtained with the SRF method was less contaminated than with the PQ method. The current FFT analysis is presented in Figure 11c,d for the PQ and SRF methods, respectively. It was clear that the THD of the source current with the SRF method was below 5% whereas with the PQ method, it was above the allowable limit with non-sinusoidal voltage conditions.

B. *Comparative performance analysis between SRF-based PA estimation and conventional PQ-based PA estimation under non-sinusoidal voltage conditions*

Two different PA estimation approaches were realized and a comparative analysis was presented; one was a conventional PA estimation based on the PQ concept and the second was the SRF-based PA estimation method [11,12]. A transient load condition of a sudden reactive load change was also considered. The reactive load demand was increased from 2.5 kVAr to 5 kVAr at 0.5 s. Figure 12a,b illustrates the PA estimation pattern for equal reactive power sharing based on the conventional PQ concept and the SRF concept, respectively. It was clearly observable that the PA with the PQ estimation method fluctuated and was also less than the reference value whereas the PA estimation with the SRF method followed the reference with fewer fluctuations. Both APFs with the SRF were found to be highly responsive to this sudden load change. These inconsistencies in the PA resulted in unequal reactive power sharing with the PQ concept, thus affecting the set criteria (Figure 12c). However, the PA with the SRF parameters resulted in more precise equal reactive power sharing, as depicted in Figure 12d. Table 3 illustrates a comparative analysis between the two approaches. Thus, it was clear that the SRF-based estimation of the PA involved fewer parameters for the estimation; the PA estimation errors and fluctuations were much fewer compared with the PQ method. Equal reactive power sharing between the shunt and series APF was more precise with fewer deviations from the reference (1.25 kVAR until 0.5 s and 2.5 kVAR after 0.5 s).

Table 2. Performance parameters under different source voltage and loading conditions.

Voltage Condition	Composite Load							Linear Load Only							Non-Linear Load Only						
	I	II	III	IV	V	VI	VII	I	II	III	IV	V	VI	VII	I	II	III	IV	V	VI	VII
Normal	228	227	0.5	4.0	9.9	3.4	$Q_{SH} = 3015$ $Q_{SR} = 2962$	229	227	0.4	4.4	0.4	3.3	$Q_{SH} = 2563$ $Q_{SR} = 2495$	228	228	0.5	2.22	18.5	3.6	-
Sag (20%)	180	225	0.5	3.7	9.8	3.8	$Q_{SH} = 3045$ $Q_{SR} = 2923$	182	226	0.5	3.4	0.5	3.2	$Q_{SH} = 2542$ $Q_{SR} = 2478$	180	226	0.5	3.26	18.63	3.6	-
Swell (20%)	275	229	0.5	2.2	9.0	3.3	$Q_{SH} = 3023$ $Q_{SR} = 2943$	273	232	0.3	3.2	0.4	3.4	$Q_{SH} = 2558$ $Q_{SR} = 2423$	275	227	0.5	3.55	16.78	3.7	-
Harmonics (10% each 3rd, 5th and 7th)	230	226	17.4	3.0	9.7	4.6	$Q_{SH} = 3010$ $Q_{SR} = 2941$	227	225	17.2	4.7	0.9	3.5	$Q_{SH} = 2574$ $Q_{SR} = 2456$	230	226	17.26	4.08	20.65	3.9	-

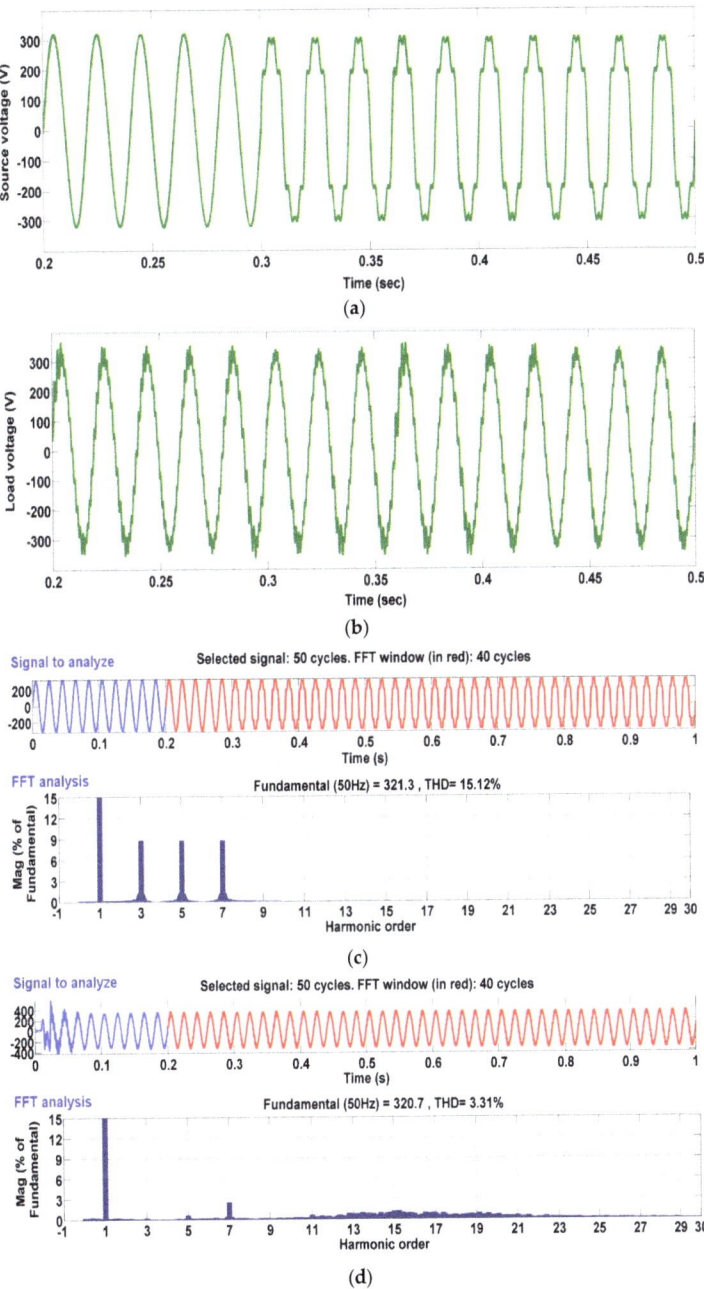

Figure 10. Voltage analysis with UPC system under harmonic conditions: (**a**) source voltage; (**b**) load voltage; (**c**) FFT analysis of source voltage; (**d**) FFT analysis of load voltage.

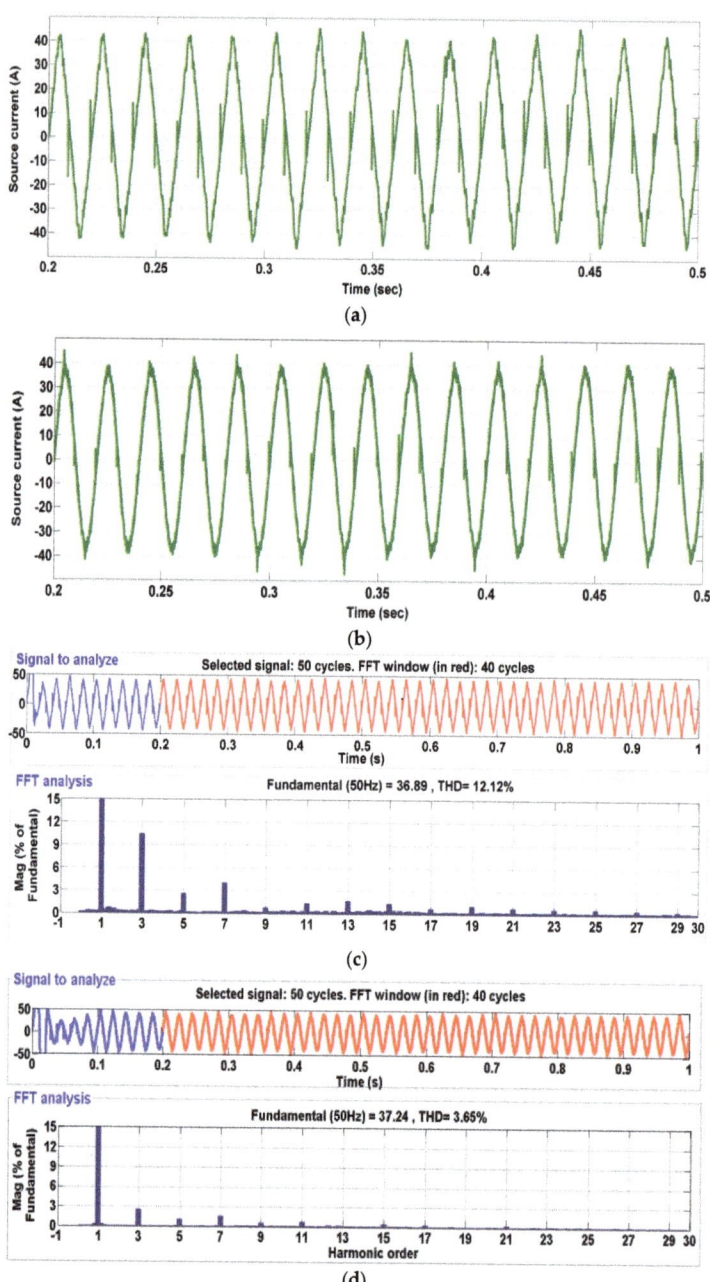

Figure 11. Source current analysis with UPC system under non-sinusoidal voltage conditions: (**a**) source current waveform with PQ method; (**b**) source current waveform with SRF method; (**c**) FFT analysis of source current with PQ method; (**d**) FFT analysis of source current with SRF method.

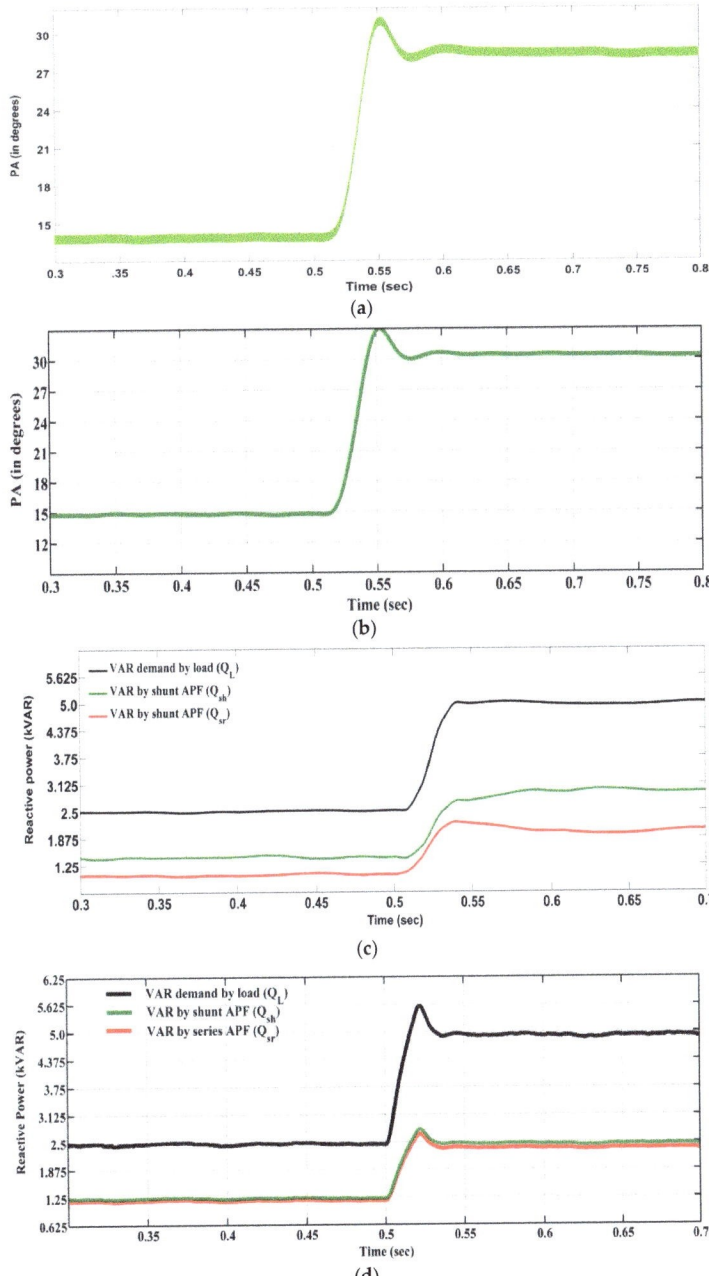

Figure 12. Comparative performance analysis between PQ and SRF methods: (**a**) PA estimation by PQ method; (**b**) PA estimation by SRF method; (**c**) reactive power share by PQ method; (**d**) reactive power share by SRF method.

Table 3. Comparative analysis between PQ and SRF methods for PA estimation.

Performance factors	PQ-based estimation of PA	SRF-based estimation of PA
Parameters involved	Load voltage and load current	Only load current
PA estimation error with reference	8%	2%
PA fluctuation with average	6.5% approximately	1.5% approximately
Percent error in equal reactive power sharing between shunt and series APF	For Q_L = 2.5 kVAR: Q_{sh} = 16%, Q_{Sr} = −17.6% For Q_L = 5 kVAR: Q_{sh} = 18%, Q_{Sr} = −20%	For Q_L = 2.5 kVAR: Q_{Sh} = 1.5%, Q_{Sr} = −2.3% For Q_L = 5 kVAR: Q_{Sh} = 1.1%, Q_{Sr} = −2.6%

7. Real-Time Simulator Analysis

Simulators are useful in practice to protect equipment from being damaged. With the reduction in cost and increments in the performance of virtual and real-time simulative technology, its availability and applicability has been widespread [14]. The real-time simulator adapted in our work was RT-LAB, which is based on FPGA; its flexibility and scalability can be used for virtually any simulation or control system application.

A computer system with installed SIMULINK software was connected to the simulator setup via an ethernet, as shown in Figure 13. Instead of methods such as HIL (hardware in the loop), RCP (rapid control prototyping) was adapted for analyzing the behavior of the UPC system with the proposed controller.

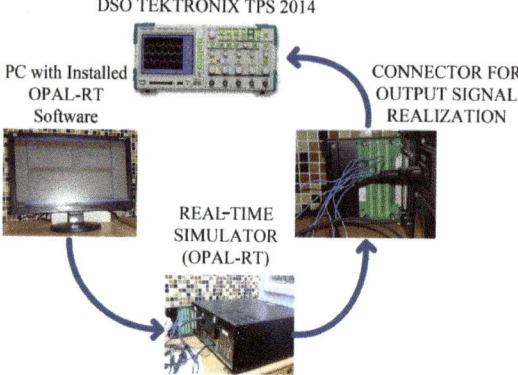

Figure 13. OPAL-RT simulator connected to the host PC via ethernet and to the DSO via connecting probes.

Figure 14 illustrates the waveforms of the source voltage, load voltage and series injected voltage for three different transient conditions of source voltage such as normal to sag (Figure 14a), sag to swell (Figure 14b) and swell condition to normal voltage with harmonics (Figure 14c). The series injected voltage compensated for the source voltage and the load voltage was found to be maintained at its rated value and free from harmonics irrespective of the disturbances. Figure 15 shows the waveforms of the load current, source current and compensating current for different load conditions of the linear load (Figure 15a), non-linear load (Figure 15b) and composite load (Figure 15c), respectively. The source current was found to be distortion-free in the case of the non-linear and composite loads. With the linear load condition, the presence of a compensating current was due to the partial load reactive current demand from the shunt APF; the remainder of the reactive current demand was fulfilled by the series APF.

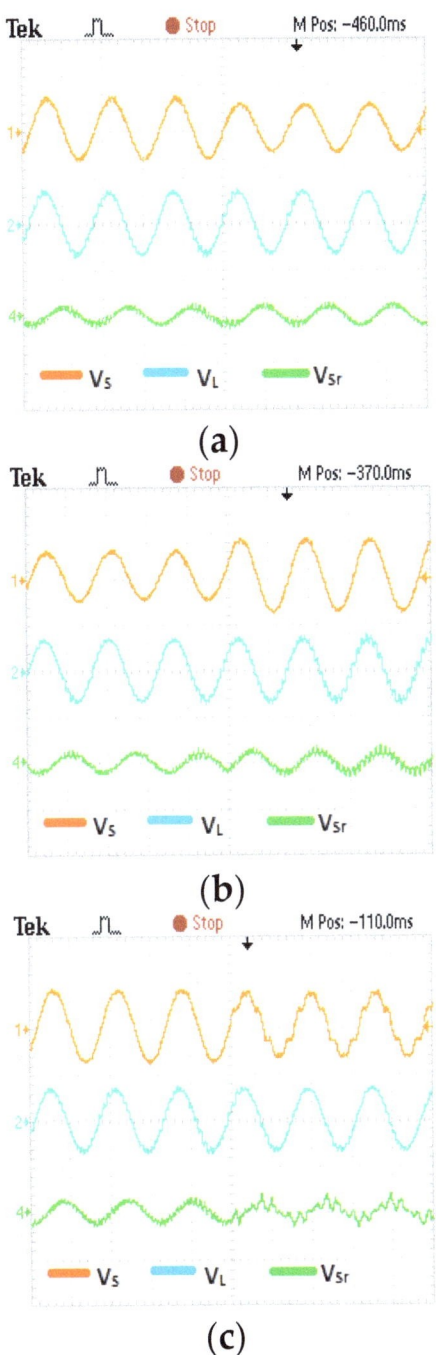

Figure 14. OPAL-RT results of source voltage (V_S), load voltage (V_L) and series injected voltage (V_{Sr}) for different disturbance conditions: (**a**) normal to sag; (**b**) sag to swell; (**c**) swell to normal with harmonics.

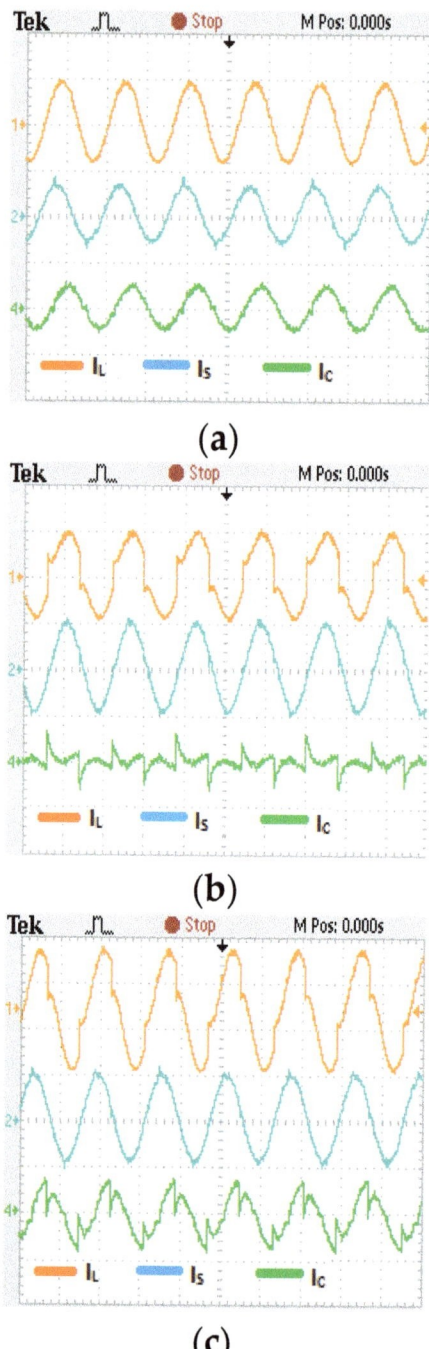

Figure 15. OPAL-RT results of load current (I_L), source current (I_S) and compensating current (I_C) for different load conditions: (**a**) linear load only; (**b**) non-linear load only; (**c**) composite load (both linear and non-linear).

8. Conclusions

An equal reactive power sharing strategy for a single-phase UPC system was proposed in this work as a more efficient and practical way of adapting identical APFs with the same rating. The SRF-based controller employed in this system for the shunt APF controller also proved to be helpful under non-sinusoidal conditions of supply voltage for a PA estimation to implement equal reactive power sharing. In this paper, a detailed performance analysis was presented for a UPC system under different supply voltages (i.e., voltage sag, swell and harmonics) and loading conditions (i.e., a non-linear load, linear load and composite load). The performance indices considered for this analysis were current harmonic compensation, load voltage compensation and reactive power compensation with equal sharing criteria between the shunt and series APF. It was clearly observed from the result analysis and tabular data compilation that the UPC system offered a significant compensation performance under different conditions of voltage and load. A comparative analysis was presented between the PQ method and the SRF method for current harmonic compensation, a PA estimation and equal reactive power sharing under a non-sinusoidal supply voltage. As deduced from this comparative analysis in terms of the results and comparative data tabulation, the SRF method proved to be superior than the PQ method under non-sinusoidal supply conditions. Thus, the SRF method is not only suitable for shunt APF control implementation, but also offers better reactive power sharing support under non-sinusoidal supply conditions between a shunt and series APF, utilizing only SRF parameters.

Author Contributions: Conceptualization, N.P., R.P. and B.S.; Data curation, N.P., R.P. and B.S.; Formal analysis, R.S. and A.E.-S.; Funding acquisition, N.P. and A.E.-S.; Investigation, R.P., R.S. and A.Y.A.; Methodology, N.P., R.P., R.S. and B.S.; Project administration, A.Y.A. and A.E.-S.; Resources, B.S., A.Y.A. and A.E.-S.; Software, N.P., R.S. and B.S.; Supervision, A.Y.A. and A.E.-S.; Validation, A.Y.A.; Writing—original draft, N.P. and R.S.; Writing—review & editing, A.E.-S. All authors have read and agreed to the published version of the manuscript.

Funding: This research received no external funding.

Institutional Review Board Statement: Not applicable.

Informed Consent Statement: Not applicable.

Data Availability Statement: Not applicable.

Conflicts of Interest: The authors declare no conflict of interest.

References

1. Verma, P.K.; Goswami, G. Power Quality Issues Associated with Smart Grid: A Review. In Proceedings of the 10th International Conference on System Modeling & Advancement in Research Trends (SMART), Moradabad, India, 10–11 December 2021.
2. Kannan, R.; Krishnan, B.; Porkumaran, K.; Prabakar, S.; Elamvazuthi, I.; Srinivasan, K. Power Quality Improvement Using UPQC for Grid Interconnected Renewable Energy Source. In Proceedings of the 8th International Conference on Intelligent and Advanced Systems (ICIAS), Kuching, Malaysia, 13–15 July 2021.
3. Verma, R.; Gawre, S.K.; Patidar, N.P. An Extensive Study on Optimization and Control Techniques for Power Quality Improvement. In Proceedings of the IEEE 2nd International Conference on Electrical Power and Energy Systems (ICEPES), Bhopal, India, 10–11 December 2021.
4. Sirigiri, D.; Das, N.; Bansal, R.C. Power Quality Issue and Mitigation Technique at High PV Penetration in Electricity Grid. In Proceedings of the 31st Australasian Universities Power Engineering Conference (AUPEC), Perth, Australia, 26–30 September 2021.
5. Patnaik, N.; Panda, A.K. Comparative analysis on a shunt active power filter with different control strategies for composite loads. In Proceedings of the IEEE TENCON, Bangkok, Thailand, 22–25 October 2014.
6. Reddy, C.R.; Prasad, A.G.; Sekhar, D.C.; Goud, B.S.; Kumari, K.; Kumar, M.D. Voltage Sag and Swell Compensation in Integrated System using Advanced UPQC. In Proceedings of the International Conference on Decision Aid Sciences and Application (DASA), Sakheer, Bahrain, 7–8 December 2021.
7. Meng, L.; Ma, L.; Zhu, W.; Yan, H.; Wang, T.; Mao, W.; He, X.; Shu, Z. Control Strategy of Single-Phase UPQC for Suppressing the Influences of Low-Frequency DC-Link Voltage Ripple. *IEEE Trans. Power Electron.* **2022**, *37*, 2113–2124. [CrossRef]
8. Ray, P.; Ray, P.K.; Dash, S.K. Power Quality Enhancement and Power Flow Analysis of a PV Integrated UPQC System in a Distribution Network. *IEEE Trans. Ind. Appl.* **2022**, *58*, 201–211. [CrossRef]

9. Alam, S.J.; Arya, S.R. Volterra LMS/F based Control Algorithm for UPQC with Multi-Objective Optimized PI Controller Gains. *IEEE J. Emerg. Sel. Top. Power Electron.* **2022**, in press. [CrossRef]
10. Agarwal, N.K.; Saxena, A.; Prakash, A.; Singh, A.; Srivastava, A.; Baluni, A. Review on Unified Power Quality Conditioner (UPQC) to mitigate power quality problems. In Proceedings of the 2nd Global Conference for Advancement in Technology (GCAT), Bangalore, India, 1–3 October 2021.
11. Khadkikar, V.; Chandra, A. A new control philosophy for a unified power quality conditioner (UPQC) to coordinate load-reactive power demand between shunt and series inverters. *IEEE Trans. Power Deliv.* **2008**, *23*, 2522–2534. [CrossRef]
12. Khadkikar, V.; Chandra, A. UPQC-S: A Novel Concept of Simultaneous Voltage Sag/Swell and Load Reactive Power Compensations Utilizing Series Inverter of UPQC. *IEEE Trans. Power Electron.* **2011**, *26*, 2414–2425. [CrossRef]
13. Arya, S.R.; Alam, S.J.; Ray, P. Control algorithm based on limit cycle oscillator-FLL for UPQC-S with optimized PI gains. *CSEE J. Power Energy Syst.* **2020**, *6*, 649–661.
14. Patnaik, N.; Panda, A.K. Performance analysis of a 3 phase 4 wire UPQC system based on PAC based SRF controller with real time digital simulation. *Int. J. Electr. Power Energy Syst.* **2015**, *74*, 212–221. [CrossRef]
15. Jeraldine Viji, A.; Victoire, T.A.A. Enhanced PLL based SRF control method for UPQC with fault protection under unbalanced load conditions. *Int. J. Electr. Power Energy Syst.* **2014**, *58*, 319–328. [CrossRef]
16. Pal, Y.; Swaroop, A.; Singh, B. A comparative analysis of different magnetics supported three phase four wire unified power quality conditioners—A simulation study. *Int. J. Electr. Power Energy Syst.* **2013**, *47*, 436–447. [CrossRef]
17. Modesto, R.A.; da Silva, S.A.O.; de Oliveira, A.A.; Bacon, V.D. A versatile unified power quality conditioner applied to three phase four wire distribution systems using a dual control strategy. *IEEE Trans. Power Electron.* **2016**, *31*, 5503–5514. [CrossRef]
18. Panda, A.K.; Patnaik, N.; Patel, R. Power quality enhancement with PAC-SRF based single phase UPQC under non-ideal source voltage. In Proceedings of the IEEE INDICON, New Delhi, India, 17–20 December 2015.
19. Garces-Gomez, Y.A.; Hoyos, F.E.; Candelo-Becerra, J.E. Classic Discrete Control Technique and 3D-SVPWM Applied to a Dual Unified Power Quality Conditioner. *Appl. Sci.* **2019**, *9*, 5087. [CrossRef]
20. Panda, A.K.; Patnaik, N. Management of reactive power sharing & power quality improvement with SRF-PAC based UPQC under unbalanced source voltage condition. *Int. J. Electr. Power Energy Syst.* **2017**, *84*, 182–194.
21. Sharma, A.; Gupta, N. GCDSC-PLL and PAC Based Control of Three-Phase Four-Wire UPQC for Power Quality Improvement. In Proceedings of the Fifth International Conference on Electrical Energy Systems (ICEES), Chennai, India, 21–22 February 2019.
22. Soares, V.; Verdelho, P.; Marques, G.D. An Instantaneous Active and Reactive Current Component Method for Active Filters. *Power Electron. IEEE Trans.* **2000**, *15*, 660–669. [CrossRef]

Article

Determining the Social, Economic, Political and Technical Factors Significant to the Success of Dynamic Wireless Charging Systems through a Process of Stakeholder Engagement

Shamala Gadgil [1,2,*], Karthikeyan Ekambaram [1], Huw Davies [1], Andrew Jones [3] and Stewart Birrell [1]

1. Research Institute for Clean Growth and Future Mobility, Coventry University, Coventry CV1 5FB, UK; sendtokarthick@gmail.com (K.E.); ac2616@coventry.ac.uk (H.D.); ad2998@coventry.ac.uk (S.B.)
2. Transport and Innovation Department, Coventry City Council, Coventry CV1 2GN, UK
3. Research Institute for Responsible Business, Economies & Society, Coventry University, Coventry CV1 5FB, UK; ac0766@coventry.ac.uk
* Correspondence: evanss42@uni.coventry.ac.uk

Citation: Gadgil, S.; Ekambaram, K.; Davies, H.; Jones, A.; Birrell, S. Determining the Social, Economic, Political and Technical Factors Significant to the Success of Dynamic Wireless Charging Systems through a Process of Stakeholder Engagement. *Energies* 2022, 15, 930. https://doi.org/10.3390/en15030930

Academic Editor: Adel El-Shahat

Received: 7 December 2021
Accepted: 25 January 2022
Published: 27 January 2022

Publisher's Note: MDPI stays neutral with regard to jurisdictional claims in published maps and institutional affiliations.

Copyright: © 2022 by the authors. Licensee MDPI, Basel, Switzerland. This article is an open access article distributed under the terms and conditions of the Creative Commons Attribution (CC BY) license (https://creativecommons.org/licenses/by/4.0/).

Abstract: Globally and regionally, there is an increasing impetus to electrify the road transport system. The diversity and complexity of the road transport system pose several challenges to electrification in sectors that have higher energy usage requirements. Electric road systems (ERS) have the potential for a balancing solution. An ERS is not only an engineering project, but it is also an innovation system that is complex and composed of multiple stakeholders, requiring an interdisciplinary means of aligning problems, relations, and solutions. This study looked to determine the political, economic, social, and technical (PEST) factors by actively engaging UK stakeholders through online in-depth and semi-structured discussions. The focus is on dynamic wireless power transfer (DWPT) due to its wider market reach and on the basis that a comprehensive review of the literature indicated that the current focus is on the technical challenges and hence there is a gap in the knowledge around application requirements, which is necessary if society is to achieve its goals of electrification and GHG reduction. Qualitative analysis was undertaken to identify factors that are critical to the success of a DWPT system. The outcome of this study is knowledge of the factors that determine the function and market acceptance of DWPT. These factors can be grouped into six categories: vehicle, journey, infrastructure, economic, traffic and behaviour. These factors, the associated probability distributions attributable to these factors and the relations between them (logic functions), will form the basis for decision making when implementing DWPT as part of the wider UK electric vehicle charging infrastructure and hence support the ambition to electrify all road transport. The results will make a significant contribution to the emerging knowledge base on ERS and specifically DWPT.

Keywords: dynamic wireless power transfer; EV charging infrastructure; stakeholder engagement; electric road systems; system demand

1. Introduction

There is a drive to increase the pace of electrification of the vehicle fleet in response to global and national policy objectives [1,2]. Improvements in technology, user education, new business models, etc., are all supporting the rapid transition to electric vehicle technology. The challenge is that the road transport system is far from a homogenous entity. Those sectors of the road transport ecosystem that have high energy requirements and/or high use intensity are negatively impacted by the transition to electrification—whilst battery technology is progressing rapidly, the amount of energy that can be stored is still limited compared to existing fuel types [3]. Hence, it is increasingly being recognised that innovation in the charging infrastructure will be the key enabler towards wider electrification. Electric road systems (ERSs) that allow charging on the move, thereby overcoming the inherent limitations of the battery as an energy storage medium, are one innovation.

Electric road systems is a term that covers a broad range of solutions, including catenary systems, conductive tracks and inductive tracks. Whilst there is an advantage to these systems, their role in a transport system is unclear—the current focus of the research activity is responding to the technical challenges and whether they supplant or complement existing solutions or establish new, yet to be identified niches is an under-researched area. What is clear is that concurrent developments in static charging are reducing vehicle downtime during recharging, whilst developments in battery technology and a reduction in cost are both contributing to improvements in vehicle uptime, thereby negating the need for ERS or, more appropriately, limiting its market potential. To be successful, an ERS will need to provide a valuable contribution to the current vehicle charging ecosystem. This value will be based not only on the technology, but on how that technology meets the stakeholder requirements across a broad range of criteria that may include targeting a specific cost-point, increasing convenience or meeting other, yet to be determined, utility functions.

The purpose of the research reported in this paper is to determine the factors that would contribute to the success of an ERS, specifically a dynamic inductive charging system from a social-technical perspective—as these systems have a wider market reach compared to alternative ERS solutions [4] and is an under-researched area based on the literature. The approach adopted to achieve this was to engage with the stakeholder community within the UK. The following sections provide: a background to the challenge associated with road transport electrification and a basis for the research question (Sections 1 and 2); the approach adopted in this research (Section 3); the results (Section 4); and the interpretation of those results and what they may mean for inductive charging system development (Section 5). This is the first stakeholder engagement of its kind undertaken in the UK that brings together viewpoints of a veritable and diverse group of attendees.

2. Background

Climate change is defined as the long-term alteration of temperature and weather patterns. There is robust evidence which indicates a consistent relationship between the cumulative greenhouse gas (GHG) emissions and projected increase in global temperature of between 1.5 °C and 2 °C above pre-industrial levels by the year 2100 [5]. Regional and local impacts of global warming are already seen as a consequence of the increase in GHG emissions. There is a strong concern that these impacts will worsen with stronger future climate change [6].

The Paris Agreement, which came into force in November 2016, commits developed and developing countries to keeping global warming below 2 °C and aspiring to a target of 1.5 °C [7]. According to the emissions gap report prepared by the United Nations Environment Programme (UNEP) in 2019, the total GHG emissions in 2018 amounted to 55.3 $GtCO_2e$, of which 37.5 $GtCO_2$ was attributed to CO_2 emissions from the combustion of fossil fuels [8]. Hence, in tackling climate change, current efforts are primarily focused on reducing CO_2 and cover technologies and techniques that are deployed in four main sectors, power on the supply side and industry, transportation and buildings on the demand side. However, according to the recent UNEP report referred to above, even if all unconditional nationally determined contributions (NDCs) under the Paris Agreement were implemented, we are still on course for a 3.2 °C temperature rise. Further and immediate action is therefore required to combat climate change.

Of the four sectors mentioned, it is the transport sector that presents the greatest opportunity to respond. Taking the EU-27 as an example, transport—based on 2018 figures—accounts for almost a third of all CO_2 emissions, with road transport responsible for more than two-thirds of the transport related emissions [9,10]. Further, CO_2 emissions from road transport have increased by a factor of around 1.3 between the 1990 baseline and the most recent, 2019 figures. This compares to the decrease achieved in other sectors (see Figure 1). However, transport as a sector, which is highly reliant on fossil fuels, is perhaps also one of the most challenging to decarbonise [11].

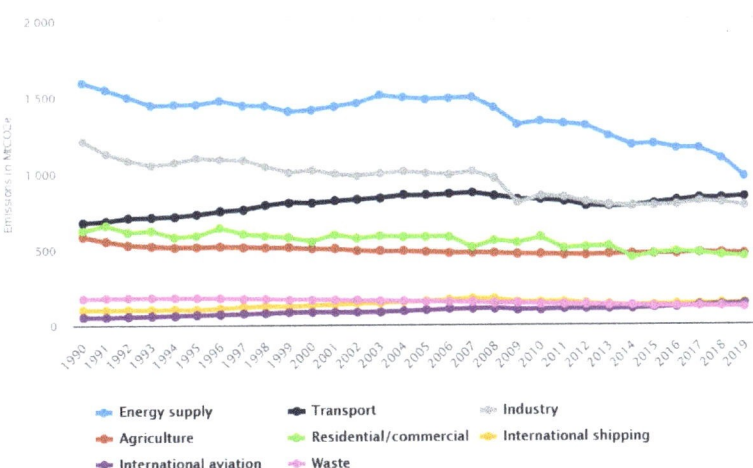

Figure 1. Annual greenhouse gas emissions in the European Union from 1990 to 2019, by sector (in million metric tons of CO_2-equivalent) [12].

The most attractive option for decarbonising the road transport sector is electrification, a market-ready technology alternative to the internal combustion engine [13]. Hence, as part of the UK commitment to net-zero GHG emissions by 2050 [14], the UK government has committed to ending "the sale of all new conventional petrol and diesel cars and vans", initially by 2040, but subsequently revised to 2035 and most recently to 2030 [15]. However, in order to achieve the net-zero goal, all road vehicles, including heavy-duty vehicles (HDVs), will need to be entirely decarbonised. Therefore, the UK Government has also announced its intention to consult on a similar phase-out to that planned for cars and vans but targeting diesel-powered heavy-goods vehicles [16].

The UK's ambitious plans to electrify its road transport fleet requires solutions that will reduce cost and drive up consumer confidence. Volume production of batteries, together with manufacturers targeting an increase in the energy density of batteries, has the potential to increase the driving range and at the same time provide a reduction in the cost of electric vehicles. The Automotive Council in the UK commissioned a roadmap on energy storage systems [17]. This roadmap targets a cost saving of around US$80 per kWh between 2017 and 2035. During the same period, it also targets the energy density to double from 250 Wh kg^{-1} to 500 Wh kg^{-1}. Further, the UK is pressing ahead with the rollout of charging infrastructure to support the electric vehicle user—both a rapid charging network to support users as they move about the UK, but also workplace and home charging solutions to support commuters and shoppers [18,19]. In addition, there is also the consideration of subsidies that either reduce cost or make available preferential access to road infrastructure including toll roads, city centres and parking [20]. However, these incentives are likely to prove unsustainable in the long term, requiring a focus on vehicle and charging technologies [21].

As energy storage costs are reducing, and technology is improving, this is encouraging OEMs to increase the battery capacity in their vehicles. What was fast charging for 16–24 kWh batteries becomes not-so-fast when the battery size reaches 40, 60 or more kWh. Further, there will also be a requirement, if we are to decarbonise all road transport sectors, to focus on public high-power charging infrastructure in support of regional and long-haul freight operations along the trunk road network. Hence, to meet the expectations of the e-mobility stakeholders and drivers, charge head providers, such as CHAdeMO, are working on higher power charging. A new edition of CHAdeMO protocol enabling 200 kW to 400 kW charging was developed by the Association and its members, who are now preparing for up to 900 kW [20]. Similar increases in charging capability are a feature of the EU's CCS system and CharIN is investigating versions up to 2 MW for electric trucks [21].

One potential challenge that the next generation of static charging technology faces is the integration of the charging stations into the existing electricity distribution grid. Recharging times are primarily becoming constrained by the electricity distribution infrastructure and not the technology of the charger/battery combination. Grid capacity is, therefore, a major issue impacting vehicle electrification and market appeal. The most obvious problem is the load increase, which can lead to a system overload since the components like transformers and supply lines are not designed to handle the extra loads requiring investment in reinforcement [22]. This leads to a juxtaposition, whereby the vehicle, through a combination of technology improvement and cost reduction has improved capability but places a requirement upon the charging system to make higher investments that need to be recouped from the system user. This tends to limit the geographical coverage of improved charging systems to where the initial investment can be recouped from higher demand or, conversely a higher operating cost for the system user.

The transition to electrification of road transport is further complicated by the diversity of different vehicle types and use cases. There is a relationship between the energy requirements of a specific modality and use case, and the ability to store energy on and transfer energy to the vehicle. Even with the development in battery technology and the reduction in cost through volume production, the difference that remains vis-à-vis fossil fuels means that replacement with electrification is not suitable for all sectors unless the frequency of charging events is substantially increased—to overcome the limitations upon on-board energy storage—and that those charging events are reduced in duration—the energy transfer rate is substantially increased in order to limit downtime, i.e., when the vehicle is stationary. As such, a transportation system based around the EV, and encompassing all possible modalities and use cases, would require the deployment of an electric refuelling infrastructure far in excess of the current fossil fuel refuelling infrastructure in order to address the driving range and recharging time limitations. These limitations have led to the discussion and exploration of various bespoke charging solutions for specific modalities and use cases, for example, the electrification of the PSV fleet in Eindhoven NL [23]

> "Recharging takes place at Hermes' depot via a wireless pantograph system on the roof of each bus. There are 20 rapid chargers for use during the day, 22 slow chargers for use during the night and 2 mobile chargers for the workshop. Alternating rapid and slow charging keeps the batteries in optimum condition. The 43 buses are used on 7 premium public transport bus line services and run at a frequency of 8 to 14 buses an hour. Eventually, 203 electric buses will be put into operation."

Another example is commercial drivers, and specifically taxi drivers, where these barriers are more pronounced as a result of the longer distances covered compared to regular cars, and fewer opportunities for recharging. Results from a recent study [24] indicated that the current plug-in charging infrastructure does not facilitate charging opportunities for taxi trade, causing longer working hours lower earnings. Drivers reported running on a range extender petrol engine once the battery is depleted, limiting the environmental benefits of electric taxis. It was concluded that alternative charger systems, including wireless, could facilitate the increased driving range of existing electric taxis by encouraging opportunistic, short but frequent charging boosts [24].

Electric road systems (ERS) are an alternative set of charging solutions that have the potential to allow electric vehicles to drive longer distances on a single at base charge, without the need to increase battery sizes or to stress the distribution grid unnecessarily. ERS have a long history and encompass in-road wireless, in-road conductive, overhead catenary, etc. A number of studies have explored the concept of the electric road, but the focus has been primarily from a technical and use case point of view [25–29]. It is considered in these studies that it is unlikely that these systems will replace existing charging systems but more likely complement (as for the PSV Eindhoven example, we create a charging ecosystem).

The most notable work concerning DWPT includes systems developed by the Korea Advanced Institute of Science and Technology (KAIST), Bombardier and Qualcomm. Other systems include the SIVETEC static WPT system developed by Siemens and the market-ready WiTricity static WPT system initially developed at Massachusetts Institute of Technology. DWPT has the potential to contribute to reducing the weight and the cost of those electric vehicles that have higher energy requirements and/or higher use intensity e.g., heavy goods vehicles or public service vehicles.

For DWPT to become a critical component of the charging ecosystem, a variety of technical challenges must be overcome. Appendix A illustrates how much existing research in the field has focused on debates surrounding technological issues. For example, Hutchinson et al. (2017) consider the technical practicalities of DWPT, whilst Gil and Taiber (2014) assessed the infrastructure challenges of introducing this technology. However, there is a lack of research into the electric road system as a whole. There is a requirement to look at the technical challenges in light of how this system is likely to be deployed and accepted by the user community.

The question that has to be asked is, what is the optimal configuration for successful integration of an electric road system as part of a wider charging ecosystem? The challenge is identifying the criteria that determine the success or limitations of DWPT and any dependencies that may exist between them. As there are multiple types of ERS, each having its associated strengths and weaknesses, the approach was to focus on the in-road dynamic wireless power transfer (DWPT) as this technology has strong adoption potential due to the ability to service multiple use cases—in-road DWPT has the potential to be deployed in multiple roadways (from urban to peri-urban through to motorway) and to service a wide range of different vehicle types (from passenger cars to public service vehicles and through to heavy goods vehicles) [4].

3. Material and Methods

The purpose of the research was to identify the factors deemed to have a significant impact on the utility and function of a DWPT system. In order to collect user and stakeholder data, a morning workshop followed by focus groups in the afternoon was organised. An approach was adopted that used the PEST framework in combination with semi-structured discussions was supported by a review of the appropriate literature in the area of DWPT and ERS. These concepts are explained below:

I. The PEST framework has been developed and proven to be successful in exploring the macro-environment [13,30]. The PEST framework (PEST referring to the political, economic, societal and technology forces present in the system) was therefore adopted as part of the research activity. PEST assesses a market (including competitors) from the viewpoint of a particular technological proposition or business. In this case the DWPT approach within the ERS is utilised as an addition to the existing market-deployed charging solutions.

II. Semi-structured discussion (that is qualitative/informal conversational/guided approach) seeks to achieve the same level of knowledge and understanding possessed by the respondent and to understand personal experiences and perceptions within a contextualised social framework [31]. In-depth and semi-structured discussions attempt to uncover underlying motives, prejudices, or attitudes towards sensitive issues.

III. A rapid review of the literature was undertaken as a time-limited and resource-efficient approach that provided relevant evidence in support of the analysis step.

By combining the PEST framework, semi-structured discussion and a rapid review of the literature, the purpose was to develop a comprehensive understanding and mapping of the macro-environment into which DWPT is to be deployed. Through this, the significant factors that would determine the success of a DWPT system could be extracted and a taxonomy or classification of the significant factors generated. The process followed a number of discrete steps that are outlined below:

3.1. Stakeholder Selection

The goal was to get the deepest possible understanding of the setting being studied. This required identification of participants who could provide information about the particular topic and setting being studied. The stakeholders were selected based on achieving a predetermined number of people from different categories—these categories were aligned to the political, economic, societal and technology forces inherent in the PEST framework. The participants were divided under four categories: Policy, Business, Consumer and Technology. In total there were 38 participants from 20 different organisations, including academia (21%), transport and highway authorities (37%), energy providers (10%), bus operators (3%), solution providers (21%) and vehicle manufacturers (8%).

3.2. Discussion

In this research activity, the adopted approach was to establish focus groups (small discussion groups) with a maximum of four stakeholders together with a group facilitator (to ask the questions and guide the process) and a rapporteur (to record the event). Focus groups enable researchers or facilitators to do most of the things they would during an interview, but with a small group. They enable a better focus on specific issues and interests and can also provide opportunities for the group to do more in-depth questioning and promote interaction—all key attributes relating to the stated research. The focus groups were preceded by the workshop where key speakers on the subject of electric vehicle charging development from a policy, economic, societal and technology perspective (one presentation on each and to all groups followed by open Q&A). The workshop introduced the current state of the art in order to empower the stakeholders with key knowledge around the subject area and beyond their immediate environs—for example, a policy-focused stakeholder will be aware of the key developments in the technology sector. In general, each of the focus group discussions lasted around 30 min. The discussions were based on the following two broad questions:

Q1. What will be the key determining factors that will support the success or failure of dynamic wireless charging as one of the charging infrastructures options?

Q2. What would be the ideal dynamic wireless charging infrastructure system from your perspective, and how would this be reached?

3.3. Transcribing

Post-event, the generated material was prepared for analysis. Qualitative (thematic) analysis of the discussion transcripts was then undertaken to seek patterns, themes and meanings that generate an in-depth understanding of the phenomenon of interest. The qualitative analysis was approached as a critical, reflective and iterative process that cycled between data and the overarching research framework that kept the big picture in mind. This approach has been applied previously for analysing semi-structured interviews [32,33]. The analysis was inherently a process of interpretation. Questions were asked of the data, informed by theory and by observations, hypotheses or hunches. If the analysis was rigorous and transparent, then the data should be able to support or not support these. This is the important part—the data needed to support or refute our ideas and should not fit the data into the story we want to tell [34,35].

3.4. Analysis

There were two parts to analysing the data, and these were as follows:
- "Content analysis" steps: read transcripts > highlight quotes and note why important > code quotes according to margin notes; and
- "Exploration analysis" steps: sort quotes into coded groups (themes) > interpret patterns in quotes > describe these patterns.

The last step in the methodology was to verify and report the results of the analysis. Verifying related to the 'reliability' (how consistent the results were) and 'validity' (whether the study investigated what was intended to be investigated) of the data. Verification was undertaken by comparing and contrasting the results of the analyses across the needs of electric vehicle charging infrastructure (as determined from the background section and supported as appropriate by the rapid review of the literature).

4. Results

The focus group discussions covered a multitude of topics, including the status of the EV, consumer behaviour, potential users, EV charging technologies, grid and transport network implications and policy and standards requisites that fit within the two broad questions. The following section reports on the core points that were discussed in the focus groups.

4.1. Policy

The section discusses the views of those stakeholders with a background in the development and/or delivery of policy, especially stakeholders who would have a future role in creating and managing the policy environment around DWPT.

Cost: Developing dynamic wireless charging infrastructure requires a high capital cost; therefore, "*an attractive business case is required*" to convince potential private or public investors who may perceive the return of investment on the technology to be uncertain, at least in the short to medium term. Since DWPT is relatively a new technology, scarcity in data availability can be a challenge for policymakers to develop business cases. Pilot studies such as DynaCov were recognised as essential to collect the necessary data. The importance of accounting for future EV landscape and user needs while developing the business case was stressed.

Infrastructure location: Selecting the right location for the infrastructure is crucial for success. It was commented that the dynamic wireless charging solutions (DWCS) must complement the other existing charging facilities and fully integrate with the charging ecosystem within the area. Furthermore, to compete with other charging solutions, the charging cost to the user must be attractive. One way to manage the cost to the user is by encouraging competition within the dynamic charging market. The road selected for the infrastructure must not possess any significant constraint for installing the infrastructure. It was noted that it might be more feasible to electrify roads within strategic road networks, including motorways and primary roads. In general, these road networks "*are well maintained and contain space for installing required utilities*". Moreover, these roads have a higher chance of carrying the required number of target vehicle types and volumes.

Temporal considerations: The disruption that may be caused during the construction phase can be detrimental to the success of the solution. It has the potential to disenfranchise the user. Therefore, installations "*must be done over a relatively short period in terms of road closures*". Furthermore, the implications of the solution for the existing road utilities and that may be installed in the future was emphasised to be evaluated. However, it was agreed that futureproofing for all needs could "*significantly increase the solution's design and the implementation cost*".

User Considerations: It was highlighted that the on-road charging solution must cater to different types of vehicles and users. However, there was a broader consensus that the DWCS will be more suited for commercial users who drive long distances and fixed routes. Private owners who usually drive fewer miles in a day are "*more likely to use home charging or destination charging facilities*". Changing consumer's behaviour was ranked as one of the main challenges for the successful adoption of this charging solution. Public engagement activities are suggested to be vital for promoting the technology. Furthermore, "*acceptance of the technology by vehicle manufacturers is essential*" to gain confidence among potential users, especially as this technology requires fixing additional components to the vehicle. Empowering the user is essential; "*the ultimate decision to use the charging facility must be with the user"*. A transparent pricing plan and a payment solution that is safe, secure and easy to use are necessary. For a better user experience, it was suggested that the payment solution should be integrated with other charging infrastructure providers within the region.

4.2. Consumer

The section discusses the views of those stakeholders that have a background in the use cases that are developed around charging infrastructure, especially stakeholders who would have a future role in determining how DWPT is likely to be used via actions that include the purchase of vehicles that are DWPT compliant.

Capital Cost (Provider): The importance of public charging facilities will remain highly relevant for the uptake of electric vehicles. The DWPT has the potential to *"relieve pressure on the number of charge heads (rapid static charging)"* that are required to meet future demand. The fear and anxiety among users to adopt a relatively new charging solution may affect the uptake of the DWCS. The users might as well think, *"if costs are coming down in 5–10 years, why invest now"*? Similarly, potential users may be concerned about the emergence of hydrogen-based fuels as a preferred solution for buses and heavy goods vehicles, which is the main target vehicle types for this technology. Therefore, stakeholders' acceptance of the product, proving the technology's longevity, and policies by the central and regional governments supporting DWCS were identified as critical factors to increase the user base.

User Behaviour: The transport pattern within the target infrastructure location drives the use of the facility. The distribution of vehicle types, user types, traffic flow, battery capacity, state of charge (SoC) thresholds, and the number of miles covered in a day by a given vehicle type are key variables that impact the charging decision-making. Similarly, the driver's range anxiety influences the decision-making for using the charging infrastructure. Smart battery management solutions are required to help make a manual or automated decision to maintain the necessary level of SoC.

Capital Cost (User): The market for the DWPT technology would be driven by commercial users who cover longer distances or consume higher levels of energy such as *"public transport services (taxies, buses and coaches), last-mile delivery and haulage companies"*. Dynamic wireless charging may also provide an option for businesses to purchase smaller electric vehicles (like the Renault Zoe), which are usually cheap. However, the range of those vehicles could be increased using this charging solution. Commercial users are generally cost-sensitive; therefore, there is a need for the solution to be cost-effective compared to other vehicle charging types. Furthermore, the business customers expect the technology to integrate easily with their system without causing any operational constraints. Dynamic wireless charging can help lower the vehicle cost by reducing the size of the battery pack. In addition, a reduction in battery size may improve the vehicles' energy consumption and utility capacity. For example, it was commented that *"for a double-decker bus, we (a large-scale national coach operator) lost about two pairs of seating capacity because of the extra space required for the batteries"*. Considering the advantages, a commercial user may opt for *"re-specifying the vehicles based on the availability of the dynamic charging facility along the routes."*

Operational Cost User: As the adoption of EVs equipped with DWPT technology grows, the challenges placed upon the power grid are most likely to increase. There is a clear interaction between the traffic pattern and demand on the grid from vehicles. Smart pricing strategies are essential to relieve the pressure on the grid during the peak power consumption period. For example, *"the cost to access the charging facility could be increased at peak times"*. However, it was also suggested that the variable pricing might not be attractive to commercial users who may prefer to know the energy cost in advance for estimating their operation cost. Therefore, an arrangement is necessary between potential largescale commercial users and the solution provider.

4.3. Business

The section discusses the views of those stakeholders that have a background in the development and/or operation of business models around DWPT, especially stakeholders who would have a future role in managing the deployment and assessment management of a DWPT solution.

User Engagement: Standardisation of the technology is vital; ensuring that the technology is compatible with different vehicle types is important to the user base. Business users are expected to be the main DWPT consumers. Therefore, the primary business model for this charging solution was predicted to be *"business-to-business rather than business-to-consumer"*. Local councils who generally operate large fleets, and public road infrastructure, transport services, freight and delivery businesses are identified as target business customers. Public engagement can help to spread knowledge and awareness about this relatively new technology. Tools such as advertisement signage and outreach campaigns can inform users about the technology and its benefits.

User Behaviour: Retrofitting the receiver and other essential components must not affect the vehicle's appearance, which is common among private vehicle owners. It was commented that the involvement of vehicle manufacturers in the designing of the receiver system installation is required. Charging the vehicle on the go may influence driving behaviour. In general, DWPTs *"incentivise travelling slowly"*; hence drivers may prefer altering their driving behaviour. Its impact on the traffic flow must be evaluated. The amount of charge a vehicle can receive depends on various external factors such as the power quality, traffic speed, and power drawn by other vehicles aligned with the same coil segment. This variability in charging may reduce user's confidence in the technology. Therefore, it is necessary to overcome this challenge with a system that could pre-inform the user about the expected amount of charging. This scenario is more relevant in an urban setting where vehicles travel bumper to bumper during peak hours.

4.4. Technology

The section discusses the views of those stakeholders with a background in the development of the underlying technology, especially stakeholders who would have a future role in supplying DWPT solutions into the marketplace.

Stakeholder Communities: Effective communication between different stakeholders, including road and grid operators, utility providers, transport authorities and local governments, is crucial for the smooth operation of the DWPT infrastructure. The development of tailored technological solutions such as an in-vehicle interface to commence and end charging, over-the-air software or firmware updates, and smart payment systems are essential for a better user experience.

Product Standards: The requirement for standards specifically designed to address technical and safety aspects of DWPT methods are vital for the successful commercialisation of the technology. Any potential health and safety concerns related to electromagnetic fields (EMF) and electromagnetic compatibility (EMC) issues associated with the technology must be addressed. For example, *"leakage of magnetic fields must not interfere with a person's health monitoring devices"*, like pacemakers. Similarly, the magnetic field generated by the system *"must not interact with communications devices"* used by emergency teams, maintenance teams and any other local key infrastructures such as airports. It is also important that when a vehicle passes over a specific segment, only that segment is activated and transmit energy to the receiver on that vehicle to avoid any risk to other nearby motorists and pedestrians using the roadway. This suggests a need for the *"development of a detailed safety case"* covering all risks, hazards and mitigation plans before deployment of the solution.

Performance Improvements: The power transfer efficiency of the system is affected by an air gap and lateral misalignment between the transmitters and receivers. It was noted that information and communication-based technology *"may be required to inform the driver to make the necessary correction"* to improve the coupling efficiency.

Performance Management: The connection of a large number of electric vehicles that arrive at different times and travel at different speeds can create varying power demand patterns upon the grid, raising power quality issues such as the injection of harmonic currents. To mitigate this, *"appropriate roadside power electronics are necessary"*. Moreover, DWCS may further load the grid during its peak demand period if it coincides with the charging facility's peak usage period. Therefore, *"future planning of the grid power supply is necessary"*. Furthermore, the impact of the DWCS on the local traffic pattern should be thoroughly evaluated.

5. Discussion

The study reconfirmed several factors, including a number of challenges that support the success and potential benefits of the DWPT system. The deployment of appropriate charging infrastructure is deemed a prerequisite for the wider uptake of EVs. Several charging infrastructure solutions are being installed across the country, such as rapid chargers, fast chargers and slow chargers [36]. They form a charging ecosystem that responds to the requirements for destination and opportunity charging. The success of the DWPT charging infrastructure will ultimately be based on the number and type of users if the initial costs are to be sufficiently amortised. The participants mentioned that the initial deployment scenarios need to maximise the usage of the technology. The commercial users who drive long distances on fixed repeatable routes are suggested as initial target users. This is supported by the literature, with Meijer [37] suggesting the deployment of DWPT systems along urban bus routes as well as short and long haul national and international freight corridors.

In order to create stakeholder acceptance of DWPT charging infrastructure, it is necessary to predict the energy demand from the EVs that potentially use the facility. From the focus group transcripts, and supported by the literature, key externalities were identified, and these externalities were classified into a taxonomy consisting of six categories:

I. Vehicle: these externalities determine whether a vehicle will make a request to charge based on factors attributable to the vehicle condition, e.g., the current SoC of the battery.
II. Journey: these externalities determine whether a user will make a request to charge based on the specific mission, e.g., during the journey, the driver can actively extend the battery range based on the journey requirement.
III. Behaviour: these externalities determine the type of charge that will result, e.g., along a fixed length of charging infrastructure, the speed of the vehicle will determine the energy transferred.
IV. Economic: these externalities determine whether a charge will be requested based on the economic considerations and alternatives, e.g., the cost to access or the cost per unit of energy.
V. Traffic: these externalities determine the demand upon the DWPT system that will result from external factors, e.g., the density and mix of traffic that flows across the system.
VI. Infrastructure: these externalities determine the broader user choice to adopt DWPT based on the availability of the system both geographically and capability, e.g., is it available on the route options for the user.

The externalities noted above are factors that will affect the uptake of DWPT either positively or negatively, and it is the combination of these factors that is important in defining demand. Figure 2 shows the taxonomy generated from the analysis of the focus-group transcripts and how the factors combine to determine demand. The structure of the taxonomy is derived from the analysis of the focus group discussions. Its relation to the published literature is discussed in the following section.

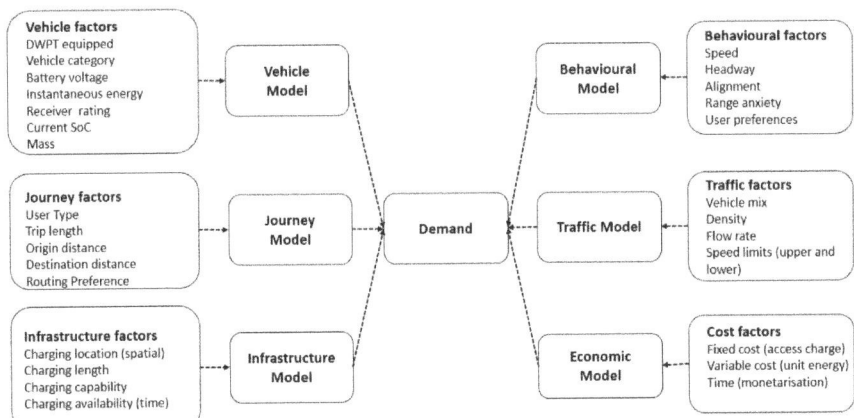

Figure 2. Taxonomy—linking Factors identified from focus groups to demand.

The focus group participants mentioned that the EV drivers are usually sensitive to costs associated with charging, concurring with previous studies [38–42]. Two concerns arose regarding cost. One was the investment cost and the recognition that, in order to mitigate the high investment cost of DWPT being passed on to the users, supportive government policies and financial subsidies, and incentives would be essential for wider uptake of the charging infrastructure, again concurring with existing studies [41]. Second was that the focus group suggested that EV users' charging behaviour is influenced by cost-driven factors such as (a) the actual cost to charge and (b) the impact of the charging method on operational efficiency. It is recognised in the literature that the actual cost to charge includes one or more components such as access fee, a kWh-based charge, usually varying with the time of use (ToU) and payment processing fees [42]. Studies have shown that smart pricing strategies based on ToU can shift charging to the off-peak period that is beneficial to the grid and, most consumers have been willing to accept this costing method. However, there is a lack of uncertainty in the willingness to use smart charging schemes between private and commercial users. Unlike static charging solutions, DWPT allows a vehicle to charge while in motion, avoiding financial losses that may be incurred due to vehicle downtime associated with stationary charging solutions. Indeed, Oliveira et al. (2020) [24] concluded that taxi drivers are more likely to lose earnings due to charging time associated with wired solutions. Furthermore, time-sensitive services like freight and public transport systems may not have enough time to get the required energy with stationary solutions [43,44]. By reducing the vehicle downtime owing to charging stops, dynamic wireless charging can be an effective solution in such scenarios [45].

Batteries constitute a significant proportion of the EV cost [46]. DWPT offers the opportunity to reduce the battery size whilst reducing vehicle cost, increasing range, and vehicle's utility (for example, capacity). Participants were also concerned about the costs associated with installing the hardware and software that are required for DWPT charging. The participants mentioned that customers expect these costs to be recouped either with a low charging price or improved operational efficiency. A similar opinion was reported in Oliveira et al. (2020) [24].

The focus group discussions suggested that the use case for the DWPT infrastructure has interdependencies with factors relating to journeys undertaken by the users, such as origin and destination of the trip, purpose of the trip and routing preference. The EV users who generally drive fewer miles than the battery range prefer destination charging at home or work [38]. Private EV users generally fall in this category. On the other hand, long-distance and energy-intensive operations such as freight and public transport services such as buses, coaches and taxis require public charging solutions at different locations.

Moreover, fixed-route services such as buses and coaches cannot detour; therefore, they rely on charging facilities along the route in addition to the infrastructure within the depots [45,47]. In general, these journey-related factors influence the willingness to alter the route for accessing a charging solution. Philipsen et al. (2015) [48] reported that users are more willing to accept a detour of 5 km or 10 min to a fast-charging station. Furthermore, Philipsen et al. (2016) [49] indicate that participants prefer to make a detour rather than accept waiting times for charging.

The participants mentioned that the design of the system needs to be interoperable, catering for a wider type of vehicle and operating conditions. However, a challenge is that there are several factors relating to vehicle characteristics as well as the technology and type of the charging system that affect the energy demand of an EV. Battery capacity is a primary factor that determines the range of an EV. A larger battery capacity enables the vehicle to travel longer distances. Therefore, when faced with a specific journey requirement, they may seek fewer charging opportunities than vehicles with smaller battery capacities. Previous research found SoC to be a key factor that determines the energy demand of an EV during a charging event. It is well recognised that specific real energy consumption that determines the SoC depends on several parameters, including battery temperature, utilisation of in-vehicle systems such as heating and ventilation, vehicle load, vehicle acceleration/deceleration, rolling friction, aerodynamic drag, and road gradient. Mishra (2018) [50] recommends operating EV batteries within a threshold range rather than taking advantage of the full range between 0% and 100%. For Li-ion batteries, the optimum SoC area is determined to be between 20% and 80%. Maintaining SoC at these levels reduces the rate of battery degradation and expands operation lifetime. Respecting the SoC range is essential; however, a driver or fleet owner may choose to operate outside this range.

A particular challenge is to develop a DWPT system capable of higher power transfer efficiencies with a wider range of lateral and vertical misalignments (air gap) between the primary source coils embedded within the road and secondary on-board coils (receiver) [51]. Naberezhnykh [52] recommends the system design to consider driver lateral misalignment of up to 15 cm for optimum usage of the charging system. The power transfer efficiency is shown to drop gradually as vertical misalignment grows [28]. Further, the air gap differs with the vehicle type, and it can vary with the loading conditions. The speed of the vehicle is expected to affect the power transfer efficiency. With an increase in the vehicle speed, the interaction time between the primary and secondary coils reduces, resulting in lower power transfer efficiency [52]. Other vehicle characteristics such as mass and vehicle length affect the energy demand of EVs. Further simulation and/or demonstrator work is required to demonstrate the relationship between the vehicle speeds and power transfer efficiencies. Heavier vehicles require more mechanical power, consuming higher energy. Vehicle type (length) determines the maximum number of vehicles in a section. It also influences the number of secondary coils (receivers) that can be fitted to the vehicle.

These findings reiterate the need for standardisation of DWPT systems and associated technologies to ensure that the deployed systems are safe, efficient and interoperable. Furthermore, standards allow manufacturers to develop and optimise their systems to the infrastructure [52]. Several organisations such as the International Organisation for Standardisation (ISO), International Electrotechnical Commission (IEC), and Society of Automotive Engineers (SAE) are currently developing standards related to DWPT, irrespective of the barriers faced due to the technical complexity and current level of maturity of the technology [53].

Studies have shown that charging behaviour is heterogeneous [54,55] among drivers. In general, factors including range anxiety, user comfort and other individual preferences contribute to charging decision making [39]. Range anxiety among EV users is a psychological barrier that can be induced by insufficient range to reach a destination or complete daily trips within a stipulated time, associated with the time required for charging and charger location. An EV driver with a higher level of range anxiety may access a charging facility when the SoC is higher than the recommended threshold. Furthermore, a higher level of range anxiety can potentially lead to dangerous driving behaviour and negatively affect drivers' emotions. As the driver's experience with the EV increases, the driver may correctly predict the EV's range concerning their range requirements, thus reducing the range anxiety [56]. The focus group participants mentioned that an in-vehicle system capable of providing a real-time charging recommendation for the driver could be helpful to overcome this challenge [57].

The participants raised issues concerning willingness to adapt driving style to suit the traffic and improve charging efficiency. The alterations to normal driving behaviour may cause inconvenience to drivers and affect their charging decision making. For example, whilst accessing the charging lane, the driver may not be able to overtake other vehicles and need to drive within a specific speed range. In addition, the driver may need to maintain distance with the leading vehicle (headway) and vehicle alignment, which can be challenging even for an experienced driver [37]. Yang and Lu (2018) [56] acknowledged that it is necessary to account for these behavioural and psychological factors for a successful mass adaptation of EV infrastructure. In those vehicles with advanced driver-assist or automated driving systems, the vehicle's alignment can be controlled by electronic systems, thus improving driver comfort. Further measures such as driver training, road markings, and lane guides can improve charging efficiency and user comfort [42]. However, further user-based research is expected to yield satisfactory measures for users to choose DWPT charging mode.

These categories and factors within the categories can be used to support the development of DWPT. Based on the above discussion, it is recognised that there will be a logical flow that will determine if a charge event takes place, the power demand and the energy that is transferred. The gateway will be the vehicle condition; the capability of the vehicle—if it is equipped with a receiver and the ability to store energy or consume energy—will determine a request to charge and the power transferred. Following this, the journey or mission of the vehicle will determine the probability of a charge event—the longer the journey, the higher energy consumption, etc., the greater the probability of a charge request. This probability will be moderated to an extent by the immediate cost–benefit of charging dynamically or statically. As described in the preceding discussion, if a charge event is required, then the choice of how to meet that charge event—static or dynamic—is a complex interplay between access costs, energy cost, monetarisation of time, etc. The behaviour of the user and the traffic environment then determines the transit time—a quicker transit time means less energy transferred, but further, a transit event that is permeated by stop-start traffic will also impact upon energy transfer. Finally, the availability of infrastructure, either geographically—leading to different vehicle routing—or having a cap based on power supply—leading to a limit on vehicles that can be serviced—will impact power levels and energy transfer.

The identified parameters, those that determine the behaviours outlined above, can be further used to estimate demand on the network using simulation modelling techniques, such as the Agent-based model (ABM) or Bayesian network model. An Agent-based model (ABM) is a class of simulation in which a system is modelled as a collection of autonomous decision-making entities called agents that interact with each other, allowing exploration of emerging behaviour of the system, usually difficult to predict in the real world [58]. Bayesian network models are structured based on Bayes' theorem, capable of updating the prior probability of some unknown variable when some evidence describing that variable exists [59].

6. Limitations

DWPT is a technology that is uncommon and rapidly evolving; therefore, identifying and inviting participants with specific expertise was difficult. While all efforts were made to invite stakeholders with expertise in different areas relating to DWPT technology and EV charging in general, the findings reported in this study may have insufficient depth in some of the topics. All the attendees were UK based; therefore, the ideas generated might not be directly transferable to the rest of the world. However, the study generated a vast amount of relevant knowledge that can form a basis for decision making when implementing this technology to support the transition to electrification within the transport sector.

7. Conclusions

In response to policy commitments, there is a requirement to electrify road transport. It is also recognised that road transport is a complex system requiring a range of solutions that include improvement at the vehicle level and innovation in charging infrastructure. One potential innovation is electric road systems (ERS), and specifically, dynamic wireless power transfer (DWPT) due to its wider market appeal relative to alternatives. Implementation of ERS requires that the challenges are identified in order that appropriate solutions can be implemented. One issue is that current techno-centric approaches do not properly consider the complex relationships between organisations, the people enacting business processes and the system that supports these processes.

This research was the first activity to focus on identifying the challenges for DWPT considering the political, economic, societal and technology perspectives and in a UK context. The research successfully brought together 38 key stakeholders and generated over 8 h of key discussion. A taxonomy of externalities—the factors that impact negatively or positively—relevant to DWPT in the UK context was generated. The taxonomy classified the externalities/factors into six categories.

- Condition of the vehicle—will it accept a charge and what power?
- Journey that is undertake—does the mission require a charge event?
- User behaviour—what will be the type of charge that will result?
- Economics—will there be a cost-benefit to a charge event?
- Level of traffic—what will be the potential energy transferred?
- Infrastructure—what is the availability for a charge event?

The definition of the factors within each category will determine if a charge event occurs and the amount of energy transferred. There will be a clear logical flow, i.e., the condition of the vehicle will be the entry point, the journey will define a need, the economics will define the choice, the infrastructure will define the energy available, the behaviour will modify that energy transfer within limits defined by the traffic. The taxonomy and logic flow will allow for creation of a systems model that will inform the decision-making process of rolling out the DWPT system. The results of this research are supported by the existing literature in this area and form the basis for decision-making when implementing DWPT as part of the wider UK electric vehicle charging infrastructure and hence support the ambition to electrify all road transport. Combined with the factors reported in this study, real-world testing will identify additional parameters that will allow the successful implementation of the DWPT system.

Author Contributions: Formal analysis: K.E., S.G. and H.D., conceptualisation: K.E., H.D. and S.G., methodology: K.E., H.D., S.G. and S.B., data curation: K.E., S.G. and H.D., validation: K.E., S.G. and H.D., resources: H.D. and S.B.; writing—original draft preparation: K.E., S.G. and H.D.; writing—review and editing: K.E., S.G., H.D., S.B. and A.J., supervision: H.D. and S.B., project administration: H.D.; funding acquisition: H.D. and S.G. All authors have read and agreed to the published version of the manuscript.

Funding: The stakeholder event was held as part of Dynamic Wireless Power Transfer, a feasibility study project, led by Coventry City Council and project partners on behalf of Western Power Distribution (WPD) using Network Innovation Allowance (NIA) funding.

Institutional Review Board Statement: The study was conducted in accordance with Coventry University's Ethical Approval process.

Informed Consent Statement: Informed consent was obtained from all subjects involved in the study.

Conflicts of Interest: The authors declare no conflict of interest.

Appendix A

Table A1. Summary of literature regarding Dynamic Wireless Power Transfer (DWPT) an electric road system (ERS).

Author	Title	Summary	Publication
Hutchinson Luke; Waterson Ben; Anvari Bani; Naberezhnykh Denis 2019	Potential of wireless power transfer for dynamic charging of electric vehicles	Paper discusses the technicalities of electric vehicles, dynamic charging infrastructure, different projects organisations developing WPT solutions along with ISOs	Journal: IET Intelligent Transport System doi:10.1049/iet-its.2018.5221
Jesko Schulte; Henrik Ny 2018	Electric road systems: Strategic stepping stone on the way towards sustainable freight transport?	The paper looks at Electric Road Systems (ERS) in comparison to the current diesel system. The Framework for Strategic Sustainable Development was used to assess whether ERS could be a stepping stone on the way towards sustainability	MDPI, Sustainability (Switzerland) doi:10.3390/su10041148
Francesco Deflorio; Luca Castello 2017	Dynamic charging-while-driving systems for freight delivery services with electric vehicles: Traffic and energy modelling	This paper develops and implements a specific traffic model based on a mesoscopic approach, where energy requirements and charging opportunities affect driving and traffic behaviours	Transportation Research Part C: Emerging Technologies doi:10.1016/j.trc.2017.04.004
Nicolaides Doros; Cebon David; Miles John 2019	An Urban Charging Infrastructure for Electric Road Freight Operations: A Case Study for Cambridge UK	This paper investigates the park and ride bus routes, the refuse collection operations, and two home delivery operations for 6 charging infrastructure for electric road freight operations	IEEE Systems Journal doi:10.1109/JSYST.2018.2864693
Chen Feng; Taylor Nathaniel; Kringos Nicole 2015	Electrification of roads: Opportunities and challenges	This paper presents the historical overview of the technology development towards the electrification of road transportation and explores in more details the Inductive Power Transfer (IPT) technology	Applied Energy doi:10.1016/j.apenergy.2015.03.067
Gil A; Taiber J 2014	A Literature Review in Dynamic Wireless Power Transfer for Electric Vehicles: Technology and Infrastructure Integration	This paper presents a literature review on the advancements of stationary and dynamic wireless power transfer used for EV charging. addressing power limitations, electromagnetic interference regulations, communication issues and interoperability, in order to point out the technology challenges to transition from stationary to dynamic wireless charging and the implementation challenges in terms of infrastructure	Springer Link, Sustainable Automotive Technologies 2013 doi:10.1007/978-3-319-01884-3_30
GPaolo Lazzeroni; Vincenzo Cirimele; Aldo Canova 2020	Economic and environmental sustainability of Dynamic Wireless Power Transfer for electric vehicles supporting reduction in local air pollutant emissions	This paper looks at the possible variations of the energy mix and the effects related to the increase in the electric energy demand related to the increase in the circulating electric vehicles.	Renewable and Sustainable Energy Reviews doi:10.1016/j.rser.2020.110537
Katsuhiro Hata; Takehiro Imura; Yoichi Hori 2016	Dynamic wireless power transfer system for electric vehicles to simplify ground facilities—power control and efficiency maximization on the secondary side	This paper looks at a novel secondary-side control method for power control and efficiency maximization. These control strategies and the controller design proposed are based on the WPT circuit analysis and the power converter model.	IEEE Applied Power Electronics Conference and Exposition (APEC) doi:10.1109/APEC.2016.7468101
Tharsis Teoh 2021	Electric vehicle charging strategies for Urban freight transport: concept and typology	This conceptual paper synthesizes the perspectives found in literature on the charging strategy concept, and provides a definition based on Orlikowski's structurational model of technology	Routledge Taylor & Francis Group doi:10.1080/01441647.2021.1950233

References

1. European Commission. EU Launches Clean Fuel Strategy. *EU Press Releases*, 4 January 2013.
2. Egbue, O.; Long, S. Barriers to Widespread Adoption of Electric Vehicles: An Analysis of Consumer Attitudes and Perceptions. *Energy Policy* **2012**, *48*, 717–729. [CrossRef]
3. Balali, Y.; Stegen, S. Review of Energy Storage Systems for Vehicles Based on Technology, Environmental Impacts, and Costs. *Renew. Sustain. Energy Rev.* **2021**, *135*, 110185. [CrossRef]

4. Schulte, J.; Ny, H. Electric Road Systems: Strategic Stepping Stone on the Way towards Sustainable Freight Transport? *Sustainability* **2018**, *10*, 1148. [CrossRef]
5. IPCC. *Climate Change 2014: Synthesis Report. Contribution*; IPCC: Geneva, Switzerland, 2014.
6. Field, C.B.; Barros, V.R. (Eds.) *Climate Change 2014—Impacts, Adaptation and Vulnerability: Regional Aspects*; Cambridge University Press: Cambridge, UK, 2014. Available online: https://www.ipcc.ch/site/assets/uploads/2018/02/ar5_wgII_spm_en.pdf (accessed on 14 July 2021).
7. Lieven, T. Policy Measures to Promote Electric Mobility—A Global Perspective. *Transp. Res. Part A Policy Pract.* **2015**, *82*, 78–93. [CrossRef]
8. UNEP. Emissions Gap Report 2019 Executive Summary. 2019. Available online: https://www.unep.org/emissions-gap-report-2020 (accessed on 14 July 2021).
9. Indicators—European Environment Agency. Available online: https://www.eea.europa.eu/data-and-maps/indicators/#c7=all&c5=transport&c0=10&b_start=0 (accessed on 28 June 2021).
10. EU Transport in Figures: Statistical Pocketbook | Eltis. Available online: https://www.eltis.org/in-brief/facts-figures/eu-transport-figures-statistical-pocketbook (accessed on 28 June 2021).
11. Neves, S.A.; Marques, A.C.; Fuinhas, J.A. Is Energy Consumption in the Transport Sector Hampering both Economic Growth and the Reduction of CO_2 Emissions? A Disaggregated Energy Consumption Analysis. *Transp. Policy* **2017**, *59*, 64–70. [CrossRef]
12. Nilsson, M.; Nykvist, B. Governing the Electric Vehicle Transition-Near Term Interventions to Support a Green Energy Economy. *Appl. Energy* **2016**, *179*, 1360–1371. [CrossRef]
13. Davies, H.; Santos, G.; Faye, I.; Kroon, R.; Weken, H. Establishing the Transferability of Best Practice in EV Policy across EU Borders. *Transp. Res. Procedia* **2016**, *14*, 2574–2583. [CrossRef]
14. UK Government. Decarbonising Transport. *Traffic Eng. Control* **2020**, *51*, 246–248.
15. UK Government. Outcome and Response to Ending the Sale of New Petrol, Diesel and Hybrid Cars and Vans. 2021. Available online: https://www.gov.uk/government/consultations/consulting-on-ending-the-sale-of-new-petrol-diesel-and-hybrid-cars-and-vans/outcome/ending-the-sale-of-new-petrol-diesel-and-hybrid-cars-and-vans-government-response (accessed on 14 July 2021).
16. Mathieu, L. Transport & Environment Further Information. How to Decarbonise the UK's Freight Sector by 2050. 2020. Available online: https://www.transportenvironment.org/publications/how-decarbonise-uks-freight-sector-2050 (accessed on 14 July 2021).
17. Technology Roadmaps. *Electrical Energy Storage Roadmap 2020*; The Advanced Propulsion Centre UK Ltd.: Coventry, UK, 2020. Available online: https://www.apcuk.co.uk/technology-%20roadmaps/ (accessed on 14 July 2021).
18. UK Government. *Workplace Charging Scheme, Guidance for Applicants, Chargepoint Installers and Manufacturers*; UK Government: London, UK, 2020; pp. 1–27.
19. Office for Low Emission Vehicles, UK Government. *Electric Vehicle Homecharge Scheme: Guidance for Customers 2016*; UK Government: London, UK, 2016.
20. High Power—Chademo Association. Available online: https://chademo.com/technology/high-power/ (accessed on 28 June 2021).
21. Kane Mark. CharIN Develops Super Powerful Charger with over 2 MW of Power. Available online: https://insideevs.com/news/372749/charin-hpccv-over-2-mw-power/ (accessed on 28 June 2021).
22. Fawzy, S.; Osman, A.I.; Doran, J.; Rooney, D.W. Strategies for Mitigation of Climate Change: A Review. *Environ. Chem. Lett.* **2020**, *18*, 2069–2094. [CrossRef]
23. Largest Emission-Free Bus Fleet in Brabant and Europe. Available online: https://www.brabantbrandbox.com/smart-mobility/electric-public-transport-buses/ (accessed on 28 June 2021).
24. Oliveira, L.; Ulahannan, A.; Knight, M.; Birrell, S. Wireless Charging of Electric Taxis: Understanding the Facilitators and Barriers to Its Introduction. *Sustainability* **2020**, *12*, 8798. [CrossRef]
25. Han, L.; Wang, S.; Zhao, D.; Li, J. The Intention to Adopt Electric Vehicles: Driven by Functional and Non-Functional Values. *Transp. Res. Part A Policy Pract.* **2017**, *103*, 185–197. [CrossRef]
26. Nicolaides, D.; Cebon, D.; Miles, J. An Urban Charging Infrastructure for Electric Road Freight Operations: A Case Study for Cambridge UK. *IEEE Syst. J.* **2019**, *13*, 2057–2068. [CrossRef]
27. Riemann, R.; Wang, D.Z.W.; Busch, F. Optimal Location of Wireless Charging Facilities for Electric Vehicles: Flow Capturing Location Model with Stochastic User Equilibrium. *Transp. Res. Part C Emerg. Technol.* **2015**, *58*, 1–12. [CrossRef]
28. Machura, P.; Li, Q. A Critical Review on Wireless Charging for Electric Vehicles. *Renew. Sustain. Energy Rev.* **2019**, *104*, 209–234. [CrossRef]
29. Choi, S.Y.; Gu, B.W.; Jeong, S.Y.; Rim, C.T. Advances in Wireless Power Transfer Systems for Roadway-Powered Electric Vehicles. *IEEE J. Emerg. Sel. Top. Power Electron.* **2015**, *3*, 18–36. [CrossRef]
30. Sandberg, A.B.; Klementsen, E.; Muller, G.; De Andres, A.; Maillet, J. Critical Factors Influencing Viability of Wave Energy Converters in Off-Grid Luxury Resorts and Small Utilities. *Sustainability* **2016**, *8*, 1274. [CrossRef]
31. Holstein, J.A. *In-Depth Interviewing*; Sage Publication: Thousand Oaks, CA, USA, 2021; pp. 103–119.
32. Zabala, A.; Sandbrook, C.; Mukherjee, N. When and How to Use Q Methodology to Understand Perspectives in Conservation Research. *Conserv. Biol.* **2018**, *32*, 1185–1194. [CrossRef]
33. Kougias, I.; Nikitas, A.; Thiel, C.; Szabó, S. Clean Energy and Transport Pathways for Islands: A Stakeholder Analysis Using Q Method. *Transp. Res. Part D Transp. Environ.* **2020**, *78*, 102180. [CrossRef]

34. Hsieh, H.F.; Shannon, S.E. Three Approaches to Qualitative Content Analysis. *Qual. Health Res.* **2005**, *15*, 1277–1288. [CrossRef]
35. Lopez Jaramillo, O.; Stotts, R.; Kelley, S.; Kuby, M. Content Analysis of Interviews with Hydrogen Fuel Cell Vehicle Drivers in Los Angeles. *Transp. Res. Rec.* **2019**, *2673*, 377–388. [CrossRef]
36. Liu, C.; Deng, K.; Li, C.; Li, J.; Li, Y.; Luo, J. The Optimal Distribution of Electric-Vehicle Chargers across a City. In Proceedings of the IEEE International Conference on Data Mining, ICDM, New Orleans, LA, USA, 18–21 November 2017; pp. 261–270. [CrossRef]
37. Amditis, A.; Karaseitanidis, G.; Damousis, I.; Guglielmi, P.; Cirimele, V. Dynamic Wireless Charging for More Efficient FEVs: The FABRIC Project Concept. In Proceedings of the MedPower 2014, Athens, Greece, 2–5 November 2014; Volume 2014. [CrossRef]
38. Jabeen, F.; Olaru, D.; Smith, B.; Braunl, T.; Speidel, S. Electric Vehicle Battery Charging Behaviour: Findings from a Driver Survey. In Proceedings of the Australasian Transport Research Forum, ATRF, Gardens Point, QLD, Australia, 2–4 October 2013.
39. Daina, N.; Sivakumar, A.; Polak, J.W. Electric Vehicle Charging Choices: Modelling and Implications for Smart Charging Services. *Transp. Res. Part C Emerg. Technol.* **2017**, *81*, 36–56. [CrossRef]
40. Ashkrof, P.; Homem de Almeida Correia, G.; van Arem, B. Analysis of the Effect of Charging Needs on Battery Electric Vehicle Drivers' Route Choice Behaviour: A Case Study in the Netherlands. *Transp. Res. Part D Transp. Environ.* **2020**, *787*, 102206. [CrossRef]
41. Heidrich, O.; Hill, A.G.; Neaimeh, M.; Huebner, Y.; Blythe, T.P.; Dawson, R.J. How Do Cities Support Electric Vehicles and What Difference Does It Make. *Technol. Forecast. Soc. Chang.* **2017**, *123*, 17–23. [CrossRef]
42. Hardman, S.; Jenn, A.; Tal, G.; Axsen, J.; Beard, G.; Daina, N.; Figenbaum, E.; Jakobsson, N.; Jochem, P.; Kinnear, N.; et al. A Review of Consumer Preferences of and Interactions with Electric Vehicle Charging Infrastructure. *Transp. Res. Part D Transp. Environ.* **2018**, *62*, 508–523. [CrossRef]
43. Shekhar, A.; Prasanth, V.; Bauer, P.; Bolech, M. Economic Viability Study of an On-Road Wireless Charging System with a Generic Driving Range Estimation Method. *Energies* **2016**, *9*, 76. [CrossRef]
44. García-Villalobos, J.; Zamora, I.; San Martín, J.I.; Asensio, F.J.; Aperribay, V. Plug-in Electric Vehicles in Electric Distribution Networks: A Review of Smart Charging Approaches. *Renew. Sustain. Energy Rev.* **2014**, *38*, 717–731. [CrossRef]
45. Deflorio, F.; Castello, L. Dynamic Charging-While-Driving Systems for Freight Delivery Services with Electric Vehicles: Traffic and Energy Modelling. *Transp. Res. Part C Emerg. Technol.* **2017**, *81*, 342–362. [CrossRef]
46. König, A.; Nicoletti, L.; Schröder, D.; Wolff, S.; Waclaw, A.; Lienkamp, M. An Overview of Parameter and Cost for Battery Electric Vehicles. *World Electr. Veh. J.* **2021**, *12*, 21. [CrossRef]
47. Liu, Z.; Song, Z. Robust Planning of Dynamic Wireless Charging Infrastructure for Battery Electric Buses. *Transp. Res. Part C Emerg. Technol.* **2017**, *83*, 77–103. [CrossRef]
48. Philipsen, R.; Schmidt, T.; Ziefle, M. A Charging Place to Be—Users' Evaluation Criteria for the Positioning of Fast-Charging Infrastructure for Electro Mobility. *Procedia Manuf.* **2015**, *3*, 2792–2799. [CrossRef]
49. Philipsen, R.; Schmidt, T.; Van Heek, J.; Ziefle, M. Fast-Charging Station Here, Please! User Criteria for Electric Vehicle Fast-Charging Locations. *Transp. Res. Part F Traffic Psychol. Behav.* **2016**, *40*, 119–129. [CrossRef]
50. Kostopoulos, E.D.; Spyropoulos, G.C.; Kaldellis, J.K. Real World Study for the Optimal Charging of Electric Vehicle. *Energy Rep.* **2020**, *6*, 418–426. [CrossRef]
51. Moon, S.; Kim, B.C.; Cho, S.Y.; Ahn, C.H.; Moon, G.W. Analysis and Design of a Wireless Power Transfer System with an Intermediate Coil for High Efficiency. *IEEE Trans. Ind. Electron.* **2014**, *61*, 5861–5870. [CrossRef]
52. Hutchinson, L.; Waterson, B.; Anvari, B.; Naberezhnykh, D. Potential of Wireless Power Transfer for Dynamic Charging of Electric Vehicles. *IET Intell. Transp. Syst.* **2019**, *3*, 3–12. [CrossRef]
53. Final Report Summary—FASTINCHARGE (Innovative Fast Inductive Charging Solution for Electric Vehicles) | Report Summary | FASTINCHARGE | FP7 | CORDIS | European Commission. Available online: https://cordis.europa.eu/project/id/314284/reporting (accessed on 28 June 2021).
54. Zoepf, S.; Mackenzie, D.; Keith, D.; Chernicoff, W. Charging Choices and Fuel Displacement in a Large-Scale Demonstration of Plug-in Hybrid Electric Vehicles. *Transp. Res. Rec.* **2013**, *2385*, 1–10. [CrossRef]
55. Azadfar, E.; Sreeram, V.; Harries, D. The Investigation of the Major Factors Influencing Plug-in Electric Vehicle Driving Patterns and Charging Behaviour. *Renew. Sustain. Energy Rev.* **2015**, *42*, 1065–1076. [CrossRef]
56. Guo, F.; Yang, J.; Lu, J. The Battery Charging Station Location Problem: Impact of Users' Range Anxiety and Distance Convenience. *Transp. Res. Part E Logist. Transp. Rev.* **2018**, *114*, 1–18. [CrossRef]
57. Sarrafan, K.; Muttaqi, K.M.; Sutanto, D.; Town, G.E. An Intelligent Driver Alerting System for Real-Time Range Indicator Embedded in Electric Vehicles. *IEEE Trans. Ind. Appl.* **2017**, *53*, 1751–1760. [CrossRef]
58. Castle, C.J.E.; Crooks, A.T. *Principles and Concepts of Agent-Based Modelling for Developing Geospatial Simulations*; Centre for Advanced Spatial Analysis, University College London (UCL): London, UK, 2006.
59. Hosseini, S. Development of a Bayesian Network Model for Optimal Site Selection of Electric Vehicle Charging Station. *Int. J. Electr. Power Energy Syst.* **2018**, *105*, 110–122. [CrossRef]

Article

Influence of Posture and Coil Position on the Safety of a WPT System While Recharging a Compact EV

Valerio De Santis [1,*], Luca Giaccone [2] and Fabio Freschi [2]

1. Department of Industrial and Information Engineering and Economics, University of L'Aquila, 67100 L'Aquila, Italy
2. Dipartimento Energia "G. Ferraris", Politecnico di Torino, Corso Duca degli Abruzzi, 24, 10129 Torino, Italy; luca.giaccone@polito.it (L.G.); fabio.freschi@polito.it (F.F.)
* Correspondence: valerio.desantis@univaq.it

Abstract: In this study, the human exposure to the magnetic field emitted by a wireless power transfer (WPT) system during the static recharging operations of a compact electric vehicle (EV) is evaluated. Specifically, the influence of the posture of realistic anatomical models, both in standing and lying positions, either inside or outside the EV, is considered. Aligned and misaligned coil configurations of the WPT system placed both in the rear and front position of the car floor are considered as well. Compliance with safety standards and guidelines has proven that reference levels are exceeded in the extreme case of a person lying on the floor with a hand close to the WPT coils, whereas the system is always compliant with the basic restrictions, at least for the considered scenarios.

Keywords: electric vehicle; EMF safety; numerical dosimetry; wireless charging; wireless power transfer

Citation: De Santis, V.; Giaccone, L.; Freschi, F. Influence of Posture and Coil Position on the Safety of a WPT System While Recharging a Compact EV. *Energies* 2021, *14*, 7248. https://doi.org/10.3390/en14217248

Academic Editor: Adel El-Shahat

Received: 8 October 2021
Accepted: 1 November 2021
Published: 3 November 2021

Publisher's Note: MDPI stays neutral with regard to jurisdictional claims in published maps and institutional affiliations.

Copyright: © 2021 by the authors. Licensee MDPI, Basel, Switzerland. This article is an open access article distributed under the terms and conditions of the Creative Commons Attribution (CC BY) license (https://creativecommons.org/licenses/by/4.0/).

1. Introduction

Due to the increasing environmental concerns, renewable energy sources have recently attracted a great deal of attention from both industry and academia [1]. A key technology following this trend is the usage of electric vehicles (EVs), whose widespread diffusion is still limited by the charging infrastructure and their on-board energy storage systems, mainly batteries [2]. To overcome so-called "range anxiety", static or dynamic wireless power transfer (WPT) systems have been proposed to recharge EVs either while they are parked or in movement [3]. However, one of the main issues related to EV-WPT systems is the large electromagnetic field (EMF) emissions during recharging operations. Indeed, the demand for fast charging has increased the power level of WPT systems from 3.3 up to 22 kW [4], yielding an EMF leakage larger than in conventional wireless systems used to recharge consumer devices. This leakage in the neighborhood environment of the car (outside and inside) has increased the need to determine the compliance of WPT systems with international safety standards and guidelines [5,6].

The exposure assessment of static and dynamic EV-WPT systems has been widely investigated [7–15]. However, while the influence of the car chassis material has been investigated in [14,15], the effect of the human posture and related positions against the WPT coils has not rigorously been addressed. Such an influence is therefore investigated in this work for a large variation of anatomical models, postures and WPT coil position/configurations. Specifically, the magnetic field emitted by a static WPT system operating at the intermediate frequency (IF) of 85 kHz and engaged in recharging the battery of a compact car, namely a FIAT 500, has been considered.

The compliance assessment of EV-WPT systems is not straightforward. Indeed, while the standalone design of the recharging system could be easily performed with classical numerical approaches, the presence of the car body, which is more difficult to take into account, has been shown to play an important role [14,15]. However, the presence of the human body does not affect the source field up to some megahertz [16,17], making it possible to separate the overall compliance procedure into two steps: (1) the simulation of the magnetic field source (WPT system and car body) and (2) numerical dosimetry (human body subject to the previously evaluated IF field). Step (1) is solved with an ad-hoc hybrid scheme coupling the boundary element method (BEM) with the surface impedance boundary conditions (SIBCs) in order to fit both the multiscale open-boundary (WPT system) and thin-sheet (car body) characteristics of the problem [18]. Step (2) is instead performed with the commercial software Sim4Life (https://zmt.swiss/sim4life, accessed on 26 October 2021), which relies on a Virtual Population (ViP). This allowed us to achieve the non-trivial task of assessing the numerical dosimetry on realistic anatomical models with different postures resembling those of a driver, of a person lying on the ground floor or in the rear-seats and of bystanders near to the car, while the WPT coils (both aligned and misaligned) were placed either in the rear or front position of the car floor due to the presence of the battery pack between the wheels.

2. Materials and Methods

2.1. Car Modeling

The compact vehicle considered in this paper is the FIAT 500, as described in [15] and freely accessible at this link https://github.com/cadema-PoliTO/vehicle4em (accessed on 26 October 2021). Once again, only the chassis of the car was considered and modeled as a surface mesh in order to exploit the capabilities of the numerical formulation, which is based on the hybrid BEM/SIBC method [18].

In contrast to [15], where the material properties of aluminum and carbon fiber were selected for the car body, in this study, a conductivity of $\sigma = 2 \times 10^6$ S/m and a relative magnetic permeability of $\mu_r = 300$ have been adopted. These values correspond to common steel with moderate shielding capabilities, as suggested in [19].

2.2. WPT System Configuration

In this paper, we have considered a WPT system classified as WPT2/Z3 by the standard SAE J2954 [4]. The input power is set to 7.7 kVA and the operational frequency is fixed to 85 kHz. For the assessment of these kinds of WPT systems, a time-harmonic formulation is sufficient because the harmonic content is negligible [19]. Furthermore, it is possible to assume a continuous sinusoidal wave even in the case that the actual waveform would be a sinusoidal burst [10].

The clearance between the receiving coil and the ground is set to 200 mm. Each coil is made of 8 turns, and the current flowing into a single turn is 26 A for the transmitter and j26 A for the receiver. Both coils are shielded by two thin layers of aluminum and ferrite with an outer dimension of approximately 420 × 420 mm^2, as shown in Figure 1.

In order to investigate the worst exposure scenario, both the case of perfect alignment (see Figure 1a,b) and of maximum misalignment, as suggested by SAE J2954 [4]—i.e., $d_x = -75$ mm and $d_y = 100$ mm (see Figure 1c,d)—were considered.

In contrast with [15], where the WPT system was placed below the car floor on the driver's side, two different locations were selected: one under the bonnet (see Figure 2) and the other under the baggage compartment (see Figure 3). This was done to avoid interference with the battery pack, which is normally placed between the rear and front wheels.

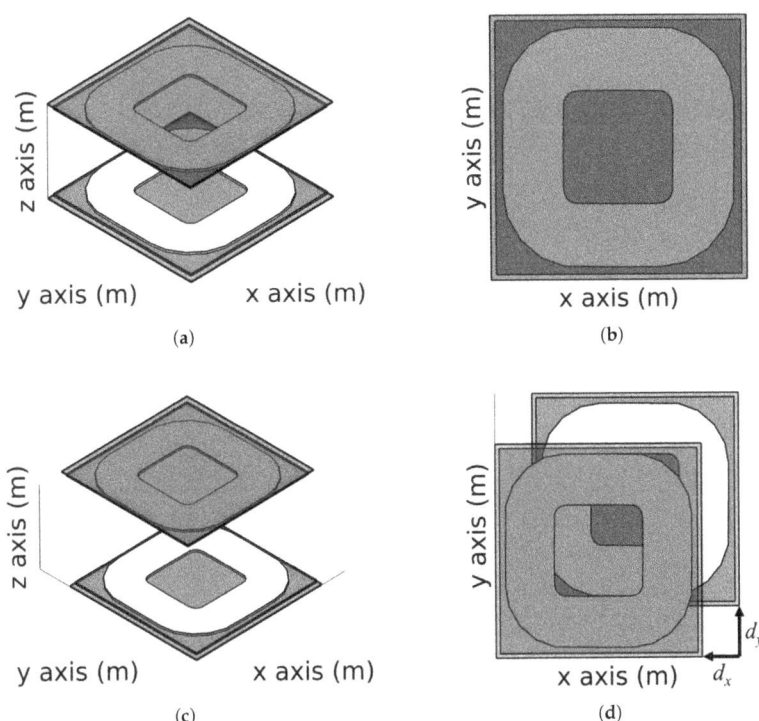

Figure 1. WPT2/Z3 system with perfect alignment (**a**,**b**) and maximum misalignment (**c**,**d**).

2.3. Exposure Scenarios for Numerical Dosimetry

Compared to [15], where a driver and a bystander were investigated (called exposure scenario #1 from here on), two further exposure scenarios, each consisting of two realistic anatomical models and two coil positions, have been considered, as illustrated in Figures 2 and 3, respectively. In particular, exposure scenario #2 consists of two adult males: one lying on the ground floor with a hand stretched towards the coils (worst-case scenario) and the other standing in front of the car. Exposure scenario #3 consists of two females: one child sleeping on the rear seats and one adult standing at the back of the car. All the anatomical human models are taken from the ViP 3.0 provided by the IT'IS Foundation (https://itis.swiss/virtual-population, accessed 26 October 2021), with a posable model for Duke lying on the floor.

Tissue dielectric properties of the human models were assigned from the IT'IS database [20], with the exception of the skin, where a higher conductivity value was adopted, as described in [21]. A uniform grid size of 2 mm was used to discretize the computational domain embedding the anatomical models (see Figures 2a and 3a).

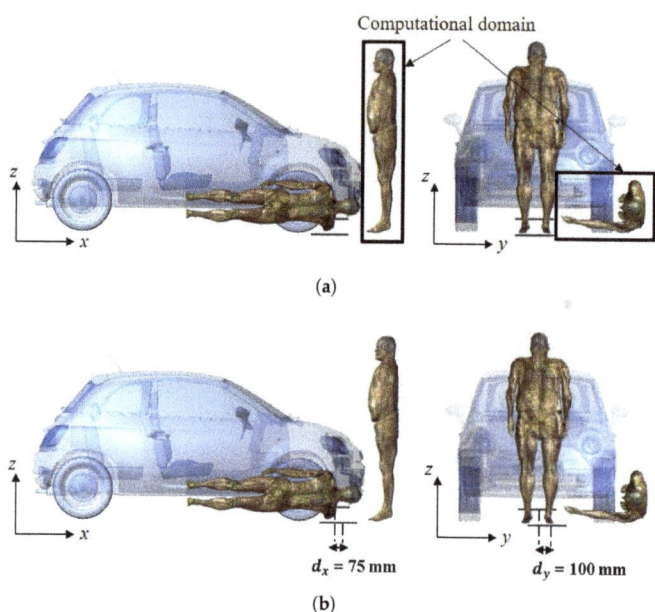

Figure 2. Exposure scenario #2: Duke lying on the ground floor and Fats standing in front of the car for the aligned (**a**) and misaligned (**b**) coils.

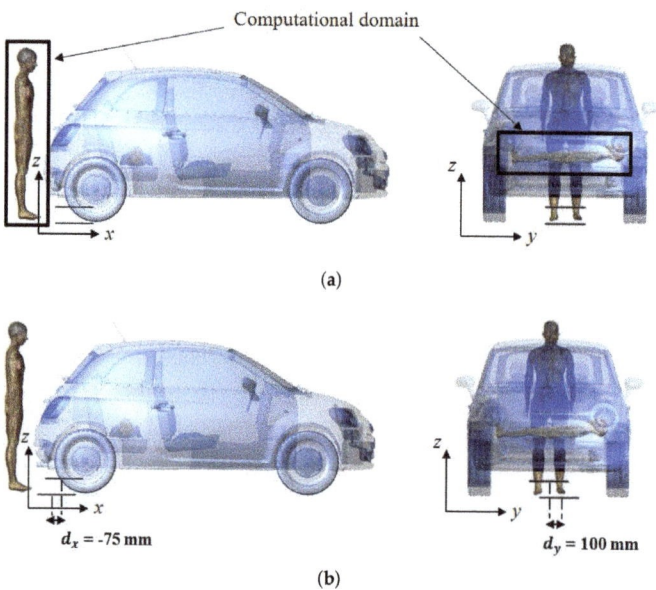

Figure 3. Exposure scenario #3: Roberta sleeping on the rear seats and Ella standing at the back of the car for the aligned (**a**) and misaligned (**b**) coils.

2.4. IF Dosimetry

It is well known that numerical dosimetry up to intermediate frequencies (10 MHz) can exploit the fact that the induced currents in the human body do not perturb the external magnetic field [16,17]. For this reason, the simulations of the external magnetic field source can be separated by the evaluation of the electric field induced inside the human bodies. This division makes it possible to select the most suitable formulations for the two steps.

In this paper, the simulation of the WPT system and the car body is handled by a numerical hybrid formulation based on BEM and SIBC methods [18]. This formulation is particularly suitable to handle the multi-scale problem as the car body has a significant surface with a very small thickness.

The numerical dosimetry computations are instead performed using the Scalar Potential Finite Element (SPFE) method, which is implemented in the commercial software tool Sim4Life. Based on the magneto-quasi-static (M-QS) approximation and the conduction-current-dominant characteristics of biological tissues in the IF region, a simplified scalar potential equation is given by

$$\nabla \cdot \sigma \nabla \phi_e = -j\omega \nabla \cdot \sigma \mathbf{A} \tag{1}$$

where $\mathbf{A}$ is the magnetic vector potential, ϕ_e is the scalar electric potential, ω is the angular frequency, and σ is the conductivity. Due to the fact that the magnetic field source is handled by a hybrid formulation based on BEM/SIBC, we cannot directly compute the necessary magnetic vector potential $\mathbf{A}$ on the right hand side of Equation (1). Therefore, the magnetic flux density $\mathbf{B}$ is computed via step (1), and a compatible magnetic vector potential $\mathbf{A}$ is then evaluated by using one of the curl-inversion procedures described in [22–24]. Specifically, Sim4Life implements the curl-inversion procedure based on Laakso et al. [22], though different schemes can be exploited by providing an external text file.

Once the magnetic vector potential $\mathbf{A}$ is provided, Equation (1) is discretized using the Galerkin Finite Element Method and linear nodal basis functions on a rectilinear grid. The resulting linear equation system is then solved using a conjugate gradient solver with a stopping criterion of 10 orders of magnitude reduction for the initial residual. Upon solving the unknown scalar potential ϕ_e, the induced electric field $\mathbf{E}$ can be computed from

$$\mathbf{E} = -\nabla \phi_e - j\omega \mathbf{A}. \tag{2}$$

3. Numerical Dosimetry Results

The aforementioned two-step approach is hereby undertaken to conduct the compliance assessment of the investigated WPT system against the EMF limits for the general public provided by the International Commission on Non-Ionizing Radiation Protection (ICNIRP) [6]. First, the $\mathbf{B}$-field is computed outside and inside the car (without the human models) by means of step (1) and compared with the reference level (RL). Then, by means of step (2), the $\mathbf{E}$-field induced inside the human body is evaluated for comparison with the basic restriction (BR).

3.1. RL Numerical Dosimetry

Figures 4 and 5 illustrate the magnetic field distributions (both aligned and misaligned coil configurations) inside the computational domains of the considered exposure scenarios #2 and #3, respectively. In these figures, the anatomical models are overlaid on the exposure scenario only for the sake of clarity; i.e., to facilitate the understanding of the compliance. As can be observed, the ICNIRP-RL is never exceeded in the sleeping Roberta model. Instead, it is barely (aligned) or moderately (misaligned) exceeded in the feet of the standing models (both front and back of the car) and is greatly exceeded in the hand (up to the wrist area) of the lying Duke model. Thus, compliance with BR is necessary only in these latter cases where the RLs are exceeded.

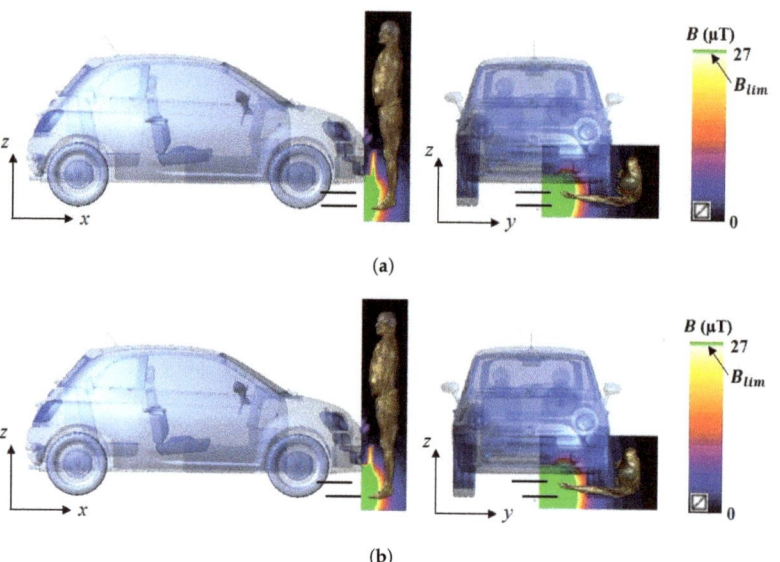

Figure 4. B-field distributions for exposure scenario #2 in the aligned (**a**) and misaligned (**b**) coil positions. B_{lim} is the RL = 27 µT (the green area is the portion where the RL is exceeded).

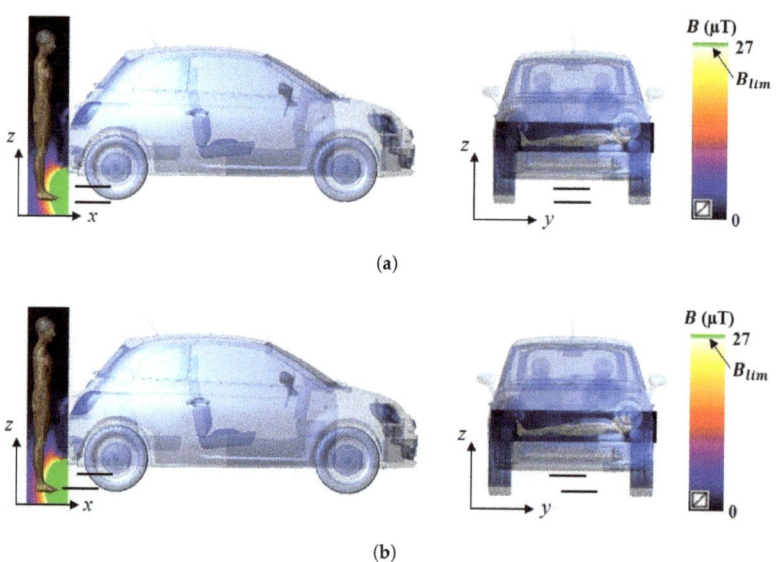

Figure 5. B-field distributions for exposure scenario #3 in the aligned (**a**) and misaligned (**b**) coil positions. B_{lim} is the RL = 27 µT (the green area is the portion where the RL is exceeded).

3.2. BR Numerical Dosimetry

The induced electric field distributions inside the different anatomical models for both exposure scenarios (#2 and #3) and both coil positions (aligned and misaligned) are reported in Figures 6 and 7, respectively. These figures show that the ICNIRP-BR is never exceeded, except for a small portion of the wrist when the lying posture on the ground floor is considered. However, it is worth noting that ICNIRP suggests determining compliance against a $2 \times 2 \times 2$ mm^3 average volume and the 99th percentile of the peak induced electric field [6]. In this work, anatomical models with a voxel resolution of 2 mm have been considered, and therefore the only 99th percentile has to be computed. Nevertheless, the 99.9th percentile is evaluated as well since the 99th percentile has sometimes been shown to underestimate the compliance, especially in the case of localized exposures [13,25–30].

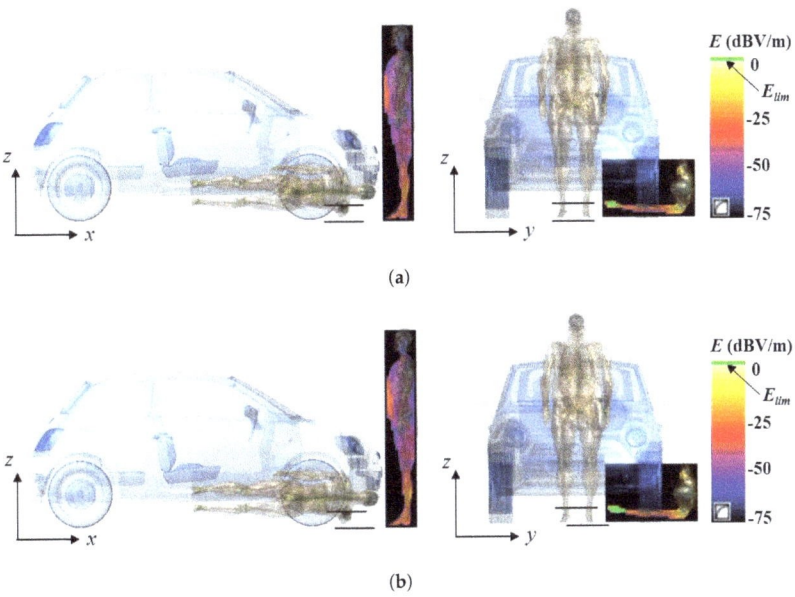

Figure 6. E-field distributions for exposure scenario #2 in the aligned (**a**) and misaligned (**b**) coil positions. E_{lim} is the BR = 11.48 V/m (the green area is the portion where the BR is exceeded).

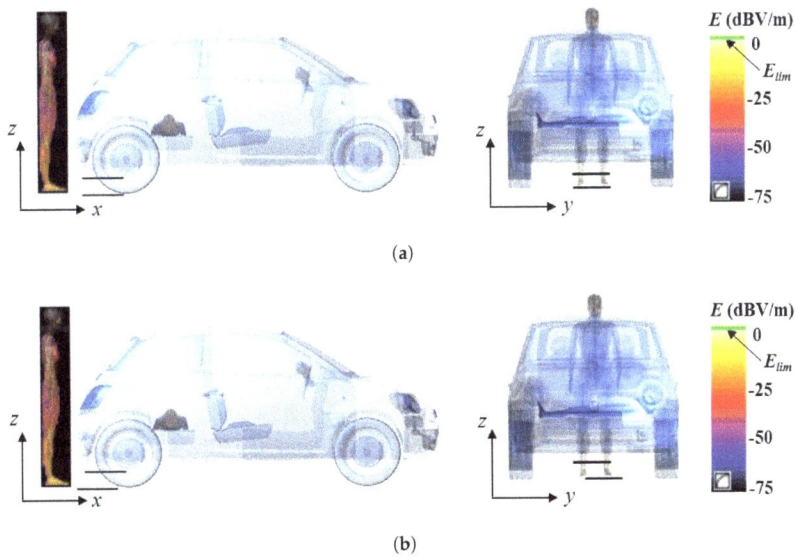

Figure 7. E-field distributions for exposure scenario #3 in the aligned (**a**) and misaligned (**b**) coil positions. E_{lim} is the BR = 11.48 V/m (the green area is the portion where the BR is exceeded).

To better quantify these results, the values of the exposure assessment are summarized in Table 1, where E_{max} is the peak induced electric field, whereas $E_{99.9}$ and E_{99} are the 99.9th and 99th percentiles, respectively. As can be observed, when comparing the latter with the BR, the overexposure is always negative (at least −6 dB), meaning that the considered exposure scenarios are far from exceeding the ICNIRP-BR.

Table 1. Summary of the compliance with the BR for the considered exposure scenarios.

Exposure Scenario	Chassis Material	Coil Position	E_{max} (V/m)	$E_{99.9}$ (V/m)	E_{99} (V/m)	Overexposure (dB)
#1 (from [15])	aluminum	Aligned	8.26	1.84	0.71	−24.17
		Misaligned	7.69	1.36	0.57	−26.08
	carbon fiber	Aligned	19.21	5.86	1.71	−16.48
		Misaligned	24.01	6.94	1.76	−16.23
#2	steel	Aligned	31.84	10.39	5.65	−6.15
		Misaligned	33.51	10.52	5.73	−6.03
#3		Aligned	1.30	0.39	0.21	−34.75
		Misaligned	2.68	0.61	0.28	−32.25

4. Conclusions and Discussions

In this paper, the influence of posture and coil position on the human safety of a WPT system engaged in recharging a compact electric vehicle was studied. The dosimetric analysis was performed by selecting a suitable mix of tools capable of analyzing the magnetic field source and evaluating the induced electric fields. The former was handled by ad-hoc software based on a hybrid scheme, whereas the latter was carried out using commercial software. This combination allowed us to handle the complex shape of the compact vehicle (namely a FIAT 500) and realistic anatomical models with different postures in a seamless way.

In order to investigate the effect of the posture and body–coil positions, a large variation of anatomical models (age, sex and body mass index) and exposure scenarios have been considered. Specifically, different postures resembling those of a driver, a lying person on the ground floor or rear-seats and bystanders near to the car were employed, while the WPT coils (both aligned and misaligned) were placed below the car floor before either the rear or front wheels due to the presence of the battery pack between the wheels.

From the analysis of the obtained results, it has been shown that the considered exposure scenarios are not compliant with the reference level, especially for a lying person with their hand close to the WPT system. Instead, compliance with the basic restriction is always satisfied, at least for the considered cases. In future, different exposure scenarios could be investigated, including heavier vehicles, such as SUVs and buses, or different anatomical models for the same exposure scenario. In the former cases, a higher power of the WPT system, together with a taller car floor, would lead to larger EMF leakages, whereas in the latter cases, different postures or anatomical details could yield higher induced fields.

Finally, it is worthy of mention that the influence of the chassis material could play a relevant role on the exposure assessment. While current steel with moderate shielding capabilities has been considered in this work, lower shielding performances have been found in previous papers by the authors when considering a futuristic chassis made of composite materials.

Author Contributions: The authors contributed equally to this work. All authors read and agreed to the published version of the manuscript.

Funding: This research received no external funding.

Institutional Review Board Statement: Not applicable.

Informed Consent Statement: Not applicable.

Data Availability Statement: Not applicable.

Acknowledgments: The authors would like to thank Donato Manesi (Politecnico di Torino) for the support given in the CAD modeling and Alessandro Franceschini (University of L'Aquila) for his valuable support with the dosimetric simulations.

Conflicts of Interest: The authors declare no conflict of interest.

References

1. IEA. Global Energy and CO_2 Status Report. 2019. Available online: www.iea.org/reports/global-energy-co2-status-report-2019 (accessed on 8 October 2021).
2. IEA. Global EV Outlook. 2020. Available online: www.iea.org/reports/global-ev-outlook-2020 (accessed on 8 October 2021).
3. Machura, P.; De Santis, V.; Li, Q. Driving Range of Electric Vehicles Charged by Wireless Power Transfer. *IEEE Trans. Veh. Technol.* **2020**, *69*, 5968–5982. [CrossRef]
4. International SAE J2954. *Wireless Power Transfer for Light-Duty Plug-In/Electric Vehicles and Alignment Methodology*; SAE International: Warrendale PA, USA, 2019. [CrossRef]
5. IEEE C95.1-2019. *IEEE Standard for Safety Levels with Respect to Human Exposure to Electric, Magnetic, and Electromagnetic Fields, 0 Hz to 300 GHz*; IEEE Press: Piscataway, NJ, USA, 2019. [CrossRef]
6. ICNIRP. Guidelines for Limiting Exposure to Time-Varying Electric and Magnetic Fields (1 Hz–100 kHz). *Health Phys.* **2010**, *99*, 818–836. [CrossRef] [PubMed]
7. Park, S.W.; Wake, K.; Watanabe, S. Incident electric field effect and numerical dosimetry for a wireless power transfer system using magnetically coupled resonances. *IEEE Trans. Microw. Theory Tech.* **2013**, *61*, 3461–3469. [CrossRef]
8. Sunohara, T.; Hirata, A.; Laakso, I.; Onishi, T. Analysis of in situ electric field and specific absorption rate in human models for wireless power transfer system with induction coupling. *Phys. Med. Biol.* **2014**, *59*, 3721. [CrossRef] [PubMed]
9. Shimamoto, T.; Laakso, I.; Hirata, A. In-situ electric field in human body model in different postures for wireless power transfer system in an electrical vehicle. *Phys. Med. Biol.* **2015**, *60*, 163–173. [CrossRef] [PubMed]
10. Cirimele, V.; Freschi, F.; Giaccone, L.; Pichon, L.; Repetto, M. Human Exposure Assessment in Dynamic Inductive Power Transfer for Automotive Applications. *IEEE Trans. Magn.* **2017**, *53*, 1–4. [CrossRef]
11. Park, S. Evaluation of electromagnetic exposure during 85 kHz wireless power transfer for electric vehicles. *IEEE Trans. Magn.* **2017**, *54*, 1–8. [CrossRef]
12. Miwa, K.; Takenaka, T.; Hirata, A. Electromagnetic dosimetry and compliance for wireless power transfer systems in vehicles. *IEEE Trans. Electromag. Compat.* **2019**, *61*, 2024–2030. [CrossRef]
13. Arduino, A.; Bottauscio, O.; Chiampi, M.; Giaccone, L.; Liorni, I.; Kuster, N.; Zilberti, L.; Zucca, M. Accuracy Assessment of Numerical Dosimetry for the Evaluation of Human Exposure to Electric Vehicle Inductive Charging Systems. *IEEE Trans. Electromag. Compat.* **2020**, *62*, 1939–1950. [CrossRef]
14. De Santis, V.; Campi, T.; Cruciani, S.; Laakso, I.; Feliziani, M. Assessment of the Induced Electric Fields in a Carbon-Fiber Electrical Vehicle Equipped with a Wireless Power Transfer System. *Energies* **2018**, *11*, 684. [CrossRef]
15. De Santis, V.; Giaccone, L.; Freschi, F. Chassis Influence on the Exposure Assessment of a Compact EV during WPT Recharging Operations. *Magnetochemistry* **2021**, *7*, 25. [CrossRef]
16. Laakso, I.; Tsuchida, S.; Hirata, A.; Kamimura, Y. Evaluation of SAR in a human body model due to wireless power transmission in the 10 MHz band. *Phys. Med. Biol.* **2012**, *57*, 4991–5002. [CrossRef] [PubMed]
17. Hirata, A.; Ito, F.; Laakso, I. Confirmation of quasi-static approximation in SAR evaluation for a wireless power transfer system. *Phys. Med. Biol.* **2013**, *58*, N241–N249. [CrossRef]
18. Freschi, F.; Giaccone, L.; Repetto, M. Algebraic formulation of nonlinear surface impedance boundary condition coupled with BEM for unstructured meshes. *Eng. Anal. Bound. Elem.* **2018**, *88*, 104–114. [CrossRef]
19. 16ENG08 EMPIR MICEV Consortium. Best Practice Guide for the Assessment of EMF Exposure from Vehicle Wireless Power Transfer Systems. 2021. Available online: www.micev.eu/ (accessed on 8 October 2021).
20. Hasgall, P.A.; Di Gennaro, F.; Baumgartner, C.; Neufeld, E.; Lloyd, B.; Gosselin, M.; Payne, D.; Klingenböck, A.; Kuster, N. IT'IS Database for Thermal and Electromagnetic Parameters of Biological Tissues. 2018. Available online: www.itis.swiss/database (accessed on 8 October 2021).
21. De Santis, V.; Chen, X.L.; Laakso, I.; Hirata, A. An equivalent skin conductivity model for low-frequency magnetic field dosimetry. *Biomed. Phys. Eng. Express* **2015**, *1*, 015201. [CrossRef]
22. Laakso, I.; De Santis, V.; Cruciani, S.; Campi, T.; Feliziani, M. Modelling of induced electric fields based on incompletely known magnetic fields. *Phys. Med. Biol.* **2017**, *62*, 6567. [CrossRef]
23. Freschi, F.; Giaccone, L.; Cirimele, V.; Canova, A. Numerical assessment of low-frequency dosimetry from sampled magnetic fields. *Phys. Med. Biol.* **2018**, *63*, 015029. [CrossRef]
24. Conchin Gubernati, A.; Freschi, F.; Giaccone, L.; Campi, T.; De Santis, V.; Laakso, I. Comparison of numerical techniques for the evaluation of human exposure from measurement data. *IEEE Trans. Magn.* **2019**, *55*, 1–4. [CrossRef]
25. Kos, B.; Valič, B.; Miklavčič, D.; Kotnik, T.; Gajšek, P. Pre- and post-natal exposure of children to EMF generated by domestic induction cookers. *Phys. Med. Biol.* **2011**, *56*, 6149–6160. [CrossRef] [PubMed]

26. Laakso, I.; Hirata, A. Reducing the staircasing error in computational dosimetry of low-frequency electromagnetic fields. *Phys. Med. Biol.* **2012**, *57*, 25–34. [CrossRef]
27. De Santis, V.; Chen, X.L. On the issues related to compliance assessment of ICNIRP 2010 basic restrictions. *J. Radiol. Prot.* **2014**, *34*, N31–N39. [CrossRef] [PubMed]
28. Schmid, G.; Hirtl, R.; Samaras, T. Dosimetric issues with simplified homogeneous body models in low frequency magnetic field exposure assessment. *J. Radiol. Prot.* **2019**, *39*, 794–808. [CrossRef] [PubMed]
29. Diao, Y.; Gomez-Tames, J.; Rashed, E.A.; Kavet, R.; Hirata, A. Spatial averaging schemes of in situ electric field for low-frequency magnetic field Eexposures. *IEEE Access* **2019**, *7*, 184320–184331. [CrossRef]
30. Giaccone, L. Compliance of non-sinusoidal or pulsed magnetic fields generated by industrial sources with reference to human exposure guidelines. In Proceedings of the 2020 International Symposium on Electromagnetic Compatibility-EMC EUROPE, Rome, Italy, 23–25 September 2020; pp. 1–6. [CrossRef]

Article

Double-Coil Dynamic Shielding Technology for Wireless Power Transmission in Electric Vehicles

Yuan Li, Shumei Zhang and Ze Cheng *

School of Electrical and Information Engineering, Tianjin University, Tianjin 300072, China; leeyuan@tju.edu.cn (Y.L.); shumeizhang@tju.edu.cn (S.Z.)
* Correspondence: chengze@tju.edu.cn; Tel.: +86-139-0201-8892

Abstract: During wireless charging, the transmission distance of electric vehicles varies, resulting in different levels of electromagnetic field leakage. An improved active shielding technology, the double-coil dynamic shielding technology, is proposed in this paper for wireless power transfer (WPT) systems with different transmission distances. Modeling, simulation, and experiments are performed for the WPT system with a double-coil dynamic shielding scheme and compared with other cases. The results show that the proposed double-coil dynamic shielding scheme is able to shield approximately 70% of the electromagnetic field leakage for WPT systems at different transmission distances. In addition, it essentially causes no degradation in transmission efficiency (only 3.1%). The effectiveness and feasibility of the proposed scheme are verified.

Keywords: electromagnetic field; wireless power transfer; shielding

1. Introduction

The promotion of electric vehicles [1] (EVs) is key to the realization of sustainable transportation. Initially, the main power transfer technologies involved battery exchange and wired charging. While battery exchange is operationally convenient, battery storage safety is still a major challenge. Wired charging, although avoiding storage safety issues, also brings some inconveniences to the utilization of EVs, such as aging wires, the risk of electric shock, and poor contact [2,3]. For this purpose, new technologies to improve comfort and safety are being investigated, of which wireless power transfer (WPT) technology is certainly one of the most appealing [4–7].

However, one of the crucial challenges of WPT systems when applied to EVs relates to the electromagnetic field (EMF) safety issues that can be caused by human exposure to severe EMFs [8–10]. During the transmission of power from the transmitting coil to the receiving coil, a portion of the EMF is radiated around the WPT system, called EMF leakage. The generation of EMF leakage not only has an impact on the devices and parts around the WPT system. It also jeopardizes the health and safety of human beings by generating current and heat inside the human body, which can cause irritation to muscles, nerves, tissues, and organs [11–13]. Therefore, the electromagnetic safety issue become an essential and critical point in the design process of WPT systems, which must ensure that the EMF leakage levels comply with the International Commission on Non-Ionizing Radiation Protection (ICNIRP) standards and guidelines [14].

Many studies have been conducted in the past few years on electromagnetic shielding technology to reduce EMF leakage from WPT systems [15–18]. Currently, there are three main shielding measures to reduce EMF leakage: passive shielding, resonant reactive current loop, and active shielding [19,20]. Features of these shielding technologies are shown in Appendix B. The present study focuses on passive shielding technology [21,22]. This suppresses electromagnetic radiation by using metallic materials to generate an EMF in the opposite direction to the one generated by the coupling coil in the form of eddy currents [23].

Despite its simplicity of operation and ease of implementation, this technology has obvious drawbacks. The use of metallic materials leads to an increase in the weight of the system, reducing the coupling coefficient and also increasing the losses [24]. Therefore, an improved passive shielding technology, resonant reactive current loop, has been proposed [25]. Its operation is based on the principle that, by placing a closed-loop coil with matching capacitance around the transmitting and receiving coils, respectively, a canceling EMF opposite to the incident field is generated to significantly reduce the EMF leakage of the WPT system. Resonant reactive current loop overcomes the shortcomings of traditional passive shielding and achieves good shielding with a small additional volume and less impact on the power transmission efficiency.

Although the method is simple in structure, the shielding coil achieves only 53% shielding effectiveness against EMF leakage, owing to the fact that the power of the shielding coil is derived from the induced EMF, yielding a limited cancellation of EMF [26]. For cases with higher shielding requirements, active shielding technology [27,28] has obvious advantages over the two technologies mentioned above. The principle of active shielding technology is to eliminate EMF leakage by generating a canceling EMF with a vector direction opposite to the incident EMF. In this technology, the active shielding coil arranged at the periphery of the transmitting coil is provided with an independent power supply, and satisfactory shielding effectiveness is achieved by adjusting the power supply.

In previous studies, a single active shielding coil around the transmitting coil was commonly taken into consideration. Therefore, the radius of the active coil should be greater than the radius of the transmitting coil [29]. With regard to the current strength and phase magnitude, the current strength should be selected based on the EMF calculation of the WPT system, while the phase should be opposite to the phase of the transmitting coil current. Therefore, the active coil generates a canceling EMF in the opposite direction of the EMF of the transmitting coil, which leads to a weakening of the EMF leakage [30]. However, the total EMF received by the receiving coil is the sum of the EMFs of the active shielding coil and transmitting coil. The algebraic sum of the two EMFs in opposite directions is significantly lower than the EMF when only the transmitting coil is present. Therefore, the addition of the active shielding coil negatively affects the transmission performance of the WPT system.

To minimize the degradation in transmission performance from the active shielding coil, multiple active coils can be considered. In this case, the geometry of each active coil, its placement, and parameters such as current strength and phase need to be discussed in more depth.

In addition, in previous WPT systems for EVs, the shielding function is usually limited to static shielding, i.e., discussing the shielding effectiveness at a fixed transmission distance [31,32]. However, for different EVs, the distance between the chassis of vehicles and the ground transmitting coil is not consistent. As a result, the power transmission distance during the charging process is inconsistent, resulting in varying EMF leakage levels.

In order to improve the wide applicability of shielding technology, this paper proposes an improved active shielding technology—the double-coil dynamic shielding scheme—for different transmission distances of EVs. Firstly, the active shielding structure with double shielding coils is discussed. Secondly, a modeling analysis of the improved, dynamically shielding WPT system is performed. Finally, simulations and experiments are performed on the proposed shielding structure to verify the effectiveness of the double-coil dynamic shielding scheme.

2. Double-Coil Active Shielding Technology

2.1. Shielding Structure

To overcome the negative impact of the addition of an active shielding coil on transmission efficiency, an improved double-coil active shielding structure is used in this paper—with two half-loops instead of a single active shielding coil—and the structure diagram is presented in Figure 1.

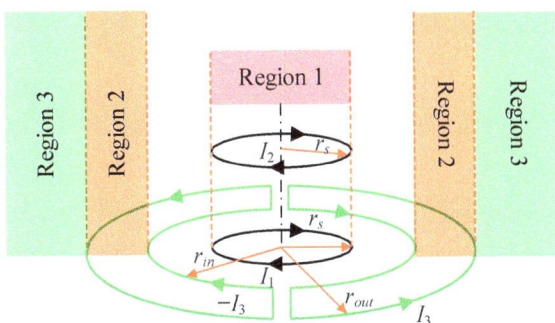

Figure 1. Double-coil active shielding structure.

It is apparent from Figure 1 that each half-loop is equivalent to half of a circle, with its inner radius denoted by r_{in} and outer radius denoted by r_{out}. Two half-loops are installed around the transmitting coil, and the radius of the transmitting coil is r_s; thus, it is obvious that $r_{in} > r_s$. To facilitate the discussion at a later stage, the regions in Figure 1 are divided as follows: inside the transmitting coil, i.e., $r < r_s$, is noted as region 1; excluding the transmitting coil part, the region inside the active shielding coil, i.e., $r_{in} < r < r_{out}$, is denoted as region 2; outside the active shielding coil, i.e., $r > r_{out}$, is noted as region 3.

Observing the direction of shielding coil currents in Figure 1, it can be seen that in region 3, the EMF generated by the two half-loops is in the opposite direction of the EMF generated by the transmitting coil, which creates a weakening effect on the total EMF in region 3. Moreover, in region 1, the EMF generated by the two shielding coils is consistent with the direction of the EMF of the transmitting coil. Compared with the traditional single shielding coil, the negative impact of the two half-loops on the performance of WPT system is reduced. It achieves a dual effect, reducing the EMF leakage of the WPT system while increasing the transmission efficiency.

2.2. Magnetic Field Calculation

In this work, the configuration described in Figure 1 is considered. Since the currents flowing in the radial direction of the two half-loops are opposite (as shown in Figure 1), the EMFs generated by these two current segments can be ignored. Consequently, in the process of calculating EMF, the structure shown in Figure 1 can be simplified. The simplified model is illustrated in Figure 2: neglecting the radial current segments, the two half-loops are replaced by two concentric circular coils with the same current but opposite phase. The inner radius and outer radius of the two concentric circular coils are consistent with Figure 1.

In Figure 2, the geometric model of the distance from the center of coil to point M is demonstrated, where the magnetic flux density at point M is a function of the coil radius, the current flowing through the coil, and the distance from the coil center to point M. The total magnetic flux density at point M is calculated by the following Equation (1) [29]:

$$B_M \approx \sum_{N=1}^{n} j\mu_0 \frac{kr_N^2 I \cos\theta_M}{2d_M^2}\left[1 + \frac{1}{jkd_M}\right]e^{-jkd_M \cdot \vartheta_N}$$
$$- \sum_{N=1}^{n} \mu_0 \frac{(kr_N)^2 I \sin\theta_M}{4d_M}\left[1 + \frac{1}{jkd_M} - \frac{1}{(kr_N)^2}\right]e^{-jkd_M} \cdot \vartheta_N \quad (1)$$

where, $d_M = \sqrt{x^2+y^2+z^2}$, $\cos\varphi_M = \frac{x}{\sqrt{x^2+y^2}}$, $\cos\theta_M = \frac{z}{\sqrt{x^2+y^2+z^2}}$, $\sin\theta_M = \frac{y}{\sqrt{x^2+y^2+z^2}}$, $\vartheta_N = \sin\theta_M \cos\varphi_M \cos\theta_M$.

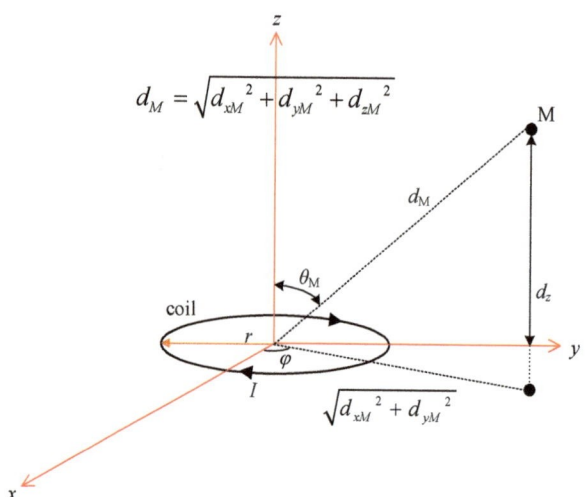

Figure 2. Geometric model of the distance from the center of the coil to point M.

Note that μ_0 is the vacuum permeability, k is the wave number ($2\pi/\lambda$), λ is the wavelength (c/f), c is the speed of light (3×10^8 m/s), f is the frequency, r_N ($N = 1, 2, \ldots, n$) is the coil radius, and N is the coil turn.

For the convenience of representation, the variable g is introduced here:

$$g = \sum_{N=1}^{n} j\mu_0 \frac{kr_N^2 \cos\theta_M}{2d_M^2}\left[1 + \frac{1}{jkd_M}\right]e^{-jkd_M \cdot \vartheta_N}$$
$$- \sum_{N=1}^{n} \mu_0 \frac{(kr_N)^2 \sin\theta_M}{4d_M}\left[1 + \frac{1}{jkd_M} - \frac{1}{(kr_N)^2}\right]e^{-jkd_M \cdot \vartheta_N} \quad (2)$$

It is abundantly clear that g is a function that varies with the position of point M(x,y,z). Therefore, Equation (1) can be simplified as (3):

$$B_M \approx g(x,y,z)I \quad (3)$$

The geometric coil array of the transmitting, receiving, and shielding coils is depicted in Figure 3. It is apparent that the transmitting, receiving, and shielding coils are all coaxial coils. Thus, all coils have an identical distance for the x-axis and y-axis to point M. The distance of the x-y axis from point M in the Cartesian coordinate system can be calculated as $\sqrt{(x^2 + y^2)}$.

However, in the z-axis direction, the distances vary from coil to coil. Therefore, the distance to point M will differ with regard to the z-axis. The distance from the transmitting coil to the point M d_{S1} is then expressed by Equation (4).

$$d_{S1} = \sqrt{\left(\sqrt{x^2 + y^2}\right)^2 + (z_1 + z_2)^2} \quad (4)$$

Since the receiving coil and shielding coil are in the same plane, the distance from the receiving coil to point M has the same expression as the distance from the shielding coil to point M, as shown in Equation (5).

$$d_{S2} = d_{SH\_in} = d_{SH\_out} = \sqrt{\left(\sqrt{x^2 + y^2}\right)^2 + z_2^2} \quad (5)$$

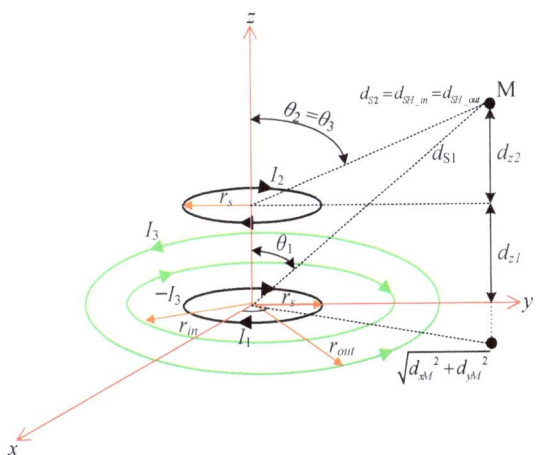

Figure 3. Geometric coil array of transmitting, receiving, and shielding coils.

where d_{S2} is the distance from the receiving coil to point M; d_{SH_in} and d_{SH_out} are, respectively, the distance from the internal and external shielding coils to point M.

The total magnetic flux density B_M at point M of the space is calculated by applying superposition as:

$$B_M = B_1 + B_2 + B_{3\_in} + B_{3\_out} \tag{6}$$

where B_1, B_2, B_{3_in}, and B_{3_out} are the magnetic flux density of the transmitting coil, receiving coil, and the internal and external shielding coils at point M, respectively, and these are calculated corresponding to the coil design through (1). Moreover, the shielding effectiveness (SE) is defined as follows:

$$SE = \left(1 - \frac{B_M}{B_1 + B_2}\right) \times 100\% \tag{7}$$

3. Dynamic Shielding for WPT Systems with Double-Coil Active Shielding

3.1. Mathematical Analysis

The theory of double-coil active coil shielding is here addressed using a circuit approach. A WPT system with a series–series (SS) compensation topology [33] is modeled using an equivalent circuit as shown in Figure 4, with R_1, R_2, and R_3 as the internal resistances; L_1, L_2, and L_3 as self-inductances; C_1, C_2, and C_3 as capacitances, and R_L and R_S as the resistive load and internal resistance of power transmitting. M_{12}, M_{13}, and M_{23} are the mutual inductances between coils, and subscript '1', '2', '3' indicates the transmitting, receiving, and shielding coil, respectively. V_S and V_A are the power supply of the transmitting coil and the shielding coil, and I_1, I_2, and I_3 are the current of the transmitting, receiving, and shielding coil, respectively. ω is the angular operating frequency.

The equivalent circuit in Figure 4 can be calculated as follows:

$$\begin{bmatrix} Z_1 + R_S & Z_{12} & Z_{13} \\ Z_{12} & Z_2 + R_L & Z_{23} \\ Z_{13} & Z_{23} & Z_3 \end{bmatrix} \begin{bmatrix} I_1 \\ I_2 \\ I_3 \end{bmatrix} = \begin{bmatrix} V_S \\ 0 \\ V_A \end{bmatrix} \tag{8}$$

$$Z_i = R_i + j\omega L_i + \frac{1}{j\omega C_i}, i \in \{1,2,3\} \tag{9}$$

$$Z_{ij} = j\omega M_{ij}, i \neq j, i,j \in \{1,2,3\} \tag{10}$$

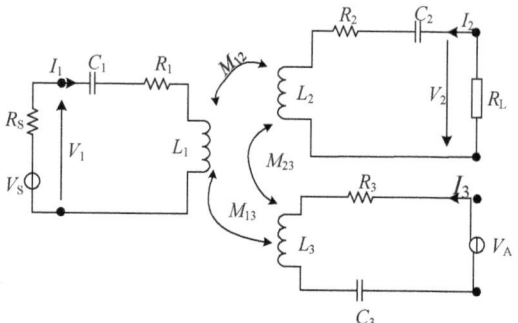

Figure 4. Equivalent circuit of WPT systems with double-coil active shielding.

It is worth mentioning that C_3 can be neglected since there is enough active current injection to compensate for the EMF leakage.

Considering, for simplicity, a three-coil configuration (1—transmitting coil, 2—receiving coil, 3—active shielding coil), the total magnetic flux density at observation point $M(x,y,z)$ in (6) can be simplified as:

$$B_M(x,y,z) = \sum_{k=1}^{3} B_k(x,y,z) \tag{11}$$

where B_3 is equal to the sum of B_{3_in} and B_{3_out}. Substituting Equation (3) into (11), there is:

$$B_M(x,y,z) = \sum_{k=1}^{3} g_k(x,y,z) I_k \tag{12}$$

It should be noted that the vectorial functions g_k depend on the configuration of all coils, i.e., the transmitting, receiving, and active shielding coils. Therefore, the feeding of the active shielding coil depends not only on the EMF generated by the transmitting coil, but also on the EMF generated by the receiving coil.

Assuming that the shielding coil achieves an ideal SE, i.e., the total magnetic flux density B_M measured at point M is zero, it yields:

$$B_3(x,y,z) = -(B_1(x,y,z) + B_2(x,y,z)) \tag{13}$$

The above Equation (13) can only be satisfied if point M lies on the surface of the plane loop with its normal axis unit vector n_a parallel to the direction of the sum of the incident fields:

$$n_a \| (g_1(x,y,z) I_1 + g_2(x,y,z) I_2) \tag{14}$$

However, in fact, for a given active coil structure, the shielding area cannot contain only one single point. Therefore, vector condition (14) cannot be completely satisfied in practice. Thus, a less restrictive but practical condition is introduced in the following.

Considering the diminishing of the main component of magnetic flux density under the condition that the active coil is properly scheduled and is planar, being parallel to the n_a, the compensation for the B_M component in the direction n_a at a given point M is given by:

$$n_a \cdot B_M(x,y,z) =$$

$$n_a \cdot \begin{bmatrix} g_1(x,y,z) & g_2(x,y,z) & g_3(x,y,z) \end{bmatrix} \cdot \begin{bmatrix} I_1 \\ I_2 \\ I_3 \end{bmatrix} = 0 \tag{15}$$

Expressing the current vector in (15) into (9), it yields:

$$n_a \cdot \begin{bmatrix} g_1(x,y,z) & g_2(x,y,z) & g_3(x,y,z) \end{bmatrix} \cdot \begin{bmatrix} Z_1+R_S & Z_{12} & Z_{13} \\ Z_{12} & Z_2+R_L & Z_{23} \\ Z_{13} & Z_{23} & Z_3 \end{bmatrix}^{-1} \begin{bmatrix} V_S \\ 0 \\ V_A \end{bmatrix} = 0 \qquad (16)$$

To make the expression more concise, a new variable t is introduced:

$$\begin{bmatrix} t_1 \\ t_2 \\ t_3 \end{bmatrix} = n_a \cdot \begin{bmatrix} g_1(x,y,z) & g_2(x,y,z) & g_3(x,y,z) \end{bmatrix} \cdot \begin{bmatrix} Z_1+R_S & Z_{12} & Z_{13} \\ Z_{12} & Z_2+R_L & Z_{23} \\ Z_{13} & Z_{23} & Z_3 \end{bmatrix}^{-1} \qquad (17)$$

Thus, Equation (16) is transformed into:

$$t_1 V_S + t_3 V_A = 0 \qquad (18)$$

From this, the expression for the power supply of the active shielding coil is further derived:

$$V_A = -V_S(t_1/t_3) \qquad (19)$$

With consideration of the losses incurred by the presence of double active shielding coils, the power transfer efficiency η of the system can be calculated as:

$$\eta = \frac{P_2}{P_1 + P_3} \qquad (20)$$

$$\begin{cases} P_1 = V_1 I_1 \\ P_2 = \dfrac{V_2^2}{R_L} \\ P_3 = V_A I_3 \end{cases} \qquad (21)$$

where P_1 and P_3 are the output power of the transmitting and active shielding coils, and P_2 is the transferred power to load.

Variations in the position or current of the transmitting and receiving coils result in a corresponding change in the magnetic flux density of the WPT system. The dynamic shielding scheme proposed in this paper is to adjust the power supply V_A in the active shielding coil according to the changes in coil position and current, allowing the excitation of the shielding coil to adapt to the changes in the EMF leakage of the WPT system.

3.2. Dynamic Shielding Scheme

As the application of EVs gradually spreads, different EVs have been presented with different structures and the distance between the vehicle chassis and the ground varies. This results in different strengths of EMF leakage when charging different EVs. If each EV needs to be designed with one active shielding mechanism, it would occupy much of the ground charging area.

Therefore, a double-coil dynamic shielding scheme based on active shielding technology is proposed in this paper. According to the power transmission distance of different EVs, the power supply of active shielding coils installed on the ground is adjusted, so that the EMF leakage of different EVs can be dynamically shielded. Regardless of the variation in the transmission distance of the WPT system, the EMF leakage level is always guaranteed to be within the safety range specified by the International Commission on Non-Ionizing Radiation Protection (ICNIRP) standards and guidelines.

With a constant charging power supply of the WPT system, the magnetic flux density varies with the transmission distance. A magnetic flux density detection module is used to detect the magnetic flux density at different transmission distances. This involves controlling the power supply V_A of the active shielding coil, so high SE of the shielding system is maintained while adapting to changes in the transmission distance, resulting in dynamic and good shielding effectiveness.

A DC-AC inverter circuit with an adjustable duty cycle is added to the control loop so as to control the power supply of the active shielding coil. By adjusting V_A, high SE for EMF leakage at varying distances is achieved in the active shielding system. The circuit design is shown in Figure 5.

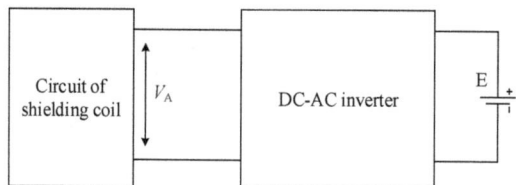

Figure 5. Circuit design of active shielding system.

A DC-AC inverter circuit is a conversion device that transforms the DC input voltage and then outputs the AC voltage. It realizes the orderly closure of switching elements by controlling the driving voltage for on/off. Thus, the high DC voltage is converted to AC output voltage according to different circulation paths. It is controlled by pulse width modulation (PWM), which further changes the magnitude of the output voltage by shifting the trigger signal of the power electronic switch in the circuit.

When the power transmission distance changes, the EMF leakage also changes accordingly. As a result, the current flowing through the coil alters as well. In the DC-AC inverter circuit, the magnitude of output voltage V_A is controlled by regulating the duty cycle of the PWM signal, so that the EMF leakage level is limited within ICNIRP.

The topologies often used in high-frequency inverter circuits are class E [34], double E [35], half-bridge, and full-bridge [36], and their circuit structures are shown in Figure 6.

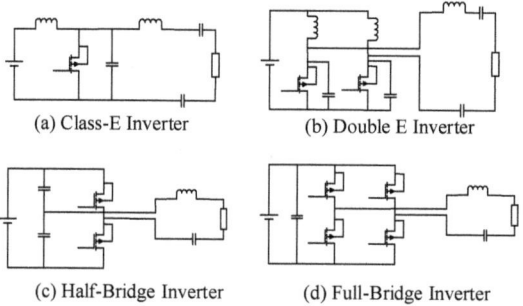

Figure 6. Circuit structures of high-frequency inverters.

A class E inverter is a simple driving circuit with only one switching device, and it has the advantages of low switching losses and high conversion efficiency. However, the circuit is unlikely to provide high output power when the duty cycle is changed.

A double E inverter consists of two switching devices, each with half the input voltage of a class E inverter. In this case, it lowers the requirement of the DC power supply and switching devices and provides an improvement in power. However, it has a large current ripple in the input inductance and a high loss in the paralleled inductor, which reduces the efficiency of the inverter system.

A half-bridge inverter works through controlling two switching devices to alternate their conduction. It features a simple structure and requires fewer switching devices. However, the maximum AC output voltage is only half of the DC output, and the lower DC voltage utilization reduces the efficiency of the inverter system.

A full-bridge inverter consists of four bridge arms, which can be seen as a combination of two half-bridge inverters. At the same DC voltage and load, the output of a full-bridge inverter is two times that of a half-bridge inverter. Moreover, it has only one capacitor on the DC side, so there is no problem of voltage balance. It can be applied in a wider range and is more flexible to control.

In summary, a full-bridge inverter is more applicable to the WPT system in this work because of its simple structure, high voltage utilization, wide power range, flexible control, and no special requirements for the transmission distance. Therefore, the full-bridge inverter is selected in this work.

With the addition of a full-bridge inverter, it is possible to adjust V_A by changing the duty cycle α of the pulse signal. In this way, the EMF leakage can be kept at a stable value while the transmission distance changes, which achieves the dynamic shielding effectiveness of the WPT system.

To further implement the dynamic shielding scheme, an MCU is added to the WPT system to guide the adjustment of α according to the detected magnetic flux density, and finally to enable the dynamic adjustment of V_A. The algorithm flowchart of the proposed dynamic shielding scheme is presented in Figure 7.

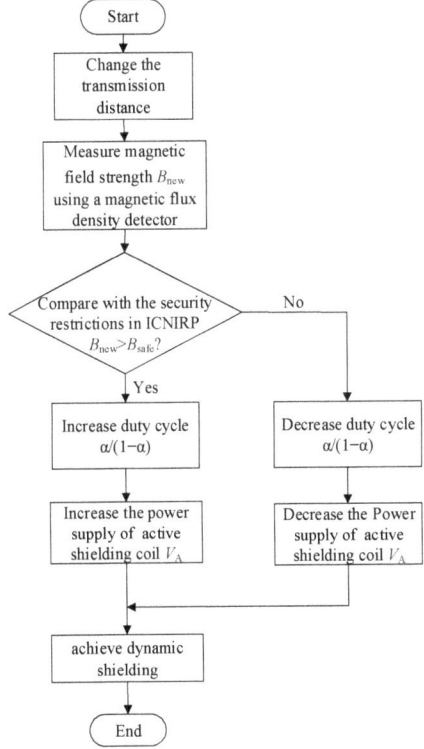

Figure 7. Flowchart of proposed dynamic shielding scheme.

The dynamic shielding scheme allows the power supply of the active shielding coils to be adjusted so that EMF leakage is restricted to the safe level of ICNIRP. The ICNIRP reference levels for magnetic field exposure for WPT systems in different operating bands (1 Hz–100 kHz) are listed in Table 1. This work refers to the ICNIRP 2010 version [37].

Table 1. Reference level of magnetic field exposure in the operating frequency range of 1 Hz–100 kHz in ICNIRP-2010.

f	B(T)
1 Hz–8 Hz	$0.04/f^2$
8 Hz–25 Hz	$0.005/f$
25 Hz–50 Hz	0.0002
50 Hz–300 Hz	0.0002
300 Hz–400 Hz	0.0002
400 Hz–3 kHz	$0.08/f$
3 kHz–100 kHz	0.000027

When the operating frequency is fixed at a certain frequency band, a standard value of the referenced magnetic flux density is fixed. However, the transmission distance may not remain the same when charging different EVs. Therefore, the magnetic flux density will increase or decrease accordingly with the distance. As the transmission distance is shortened, the EMF leakage of the WPT system will increase accordingly. To ensure that the EMF leakage is within the safe range of ICNIRP, the adoption of the proposed dynamic shielding scheme can adjust V_A accordingly. This causes the adjusted EMF leakage to drop below the reference level of ICNIRP 2010. While the transmission distance increases, the EMF leakage will become lower accordingly. From the perspective of energy saving and cost saving, V_A should be reduced accordingly, which can help to ensure that the EMF leakage is within the safety limit of the ICNIRP 2010 standard and prevent excessive waste of resources.

The proposed dynamic shielding scheme is applicable to cases in which the transmission distance changes. It has a certain directive significance for future research on shielding technology for subsequent WPT systems.

4. Simulation and Experiments

4.1. Simulation Verification

For the verification of the proposed dynamic shielding scheme based on double-coil active shielding, a WPT system is built using ANSYS Maxwell simulation software. The WPT system operates at 85 kHz, and the safety limit of ICNIRP at this operating frequency is 27 µT. The simulated structure of the WPT coils is depicted in Figure 8. The transmitting and receiving coils have the same structure, both of which have an external radius of 100 mm; the number of turns N is 10, and the transmission distance Z_2 is initialized to 100 mm. The double-coil active shielding structure consists of two half-loops with outer radius r_{out} = 150 cm and inner radius r_{in} = 130 cm.

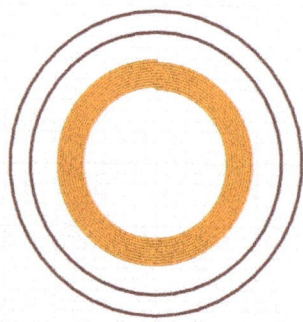

Figure 8. Simulated structure of the WPT coils.

The circuit for the combined simulation by 3D Maxwell and Simplorer is illustrated in Figure 9. The transmitting power supply V_S of the WPT system is set to 30 V. The internal resistance of power supply R_S is 8 mΩ, and the resistance of the load resistor $R_L = 10\ \Omega$. The other simulation parameters of the circuit are listed in Table 2. In addition, the variation of the power supply of the active shielding coil V_A with the variation of the distance between the transmitting coil and receiving coil is given in the subsequent discussion. In addition, the changes in the power supply of active shielding coil V_A with transmission distance are given in the subsequent discussion.

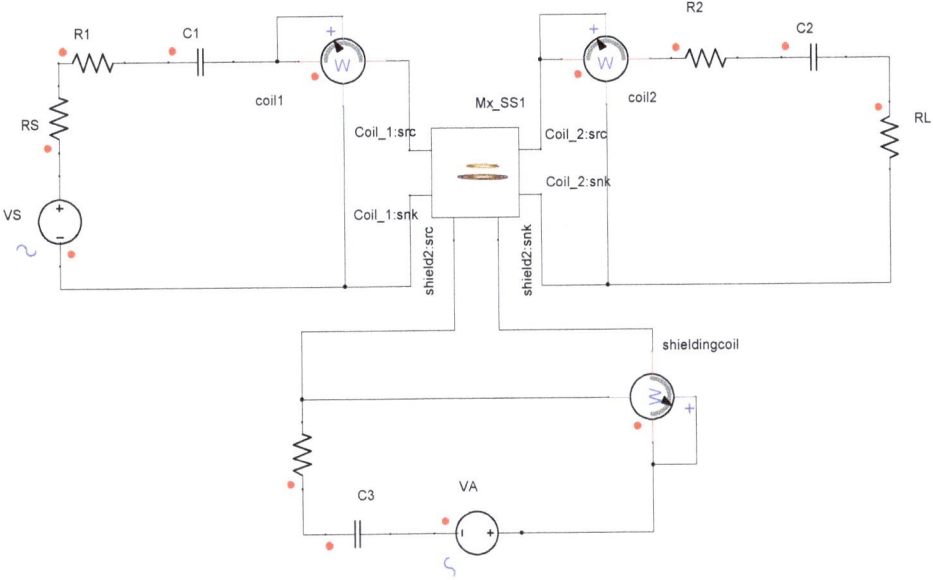

Figure 9. Circuit for the combined simulation by 3D Maxwell and Simplorer.

Table 2. Simulation parameters of WPT circuit.

Symbol	Value
R_1	15.49 mΩ
R_2	15.49 mΩ
R_3	2.64 mΩ
L_1	28.27 µH
L_2	28.27 µH
L_3	0.90 µH
C_1	4.89 µF
C_2	4.89 µF
K_{12}	0.422
K_{13}	0.041
K_{23}	0.041

In this paper, a dynamic shielding scheme is used to dynamically adjust the power supply of active shielding coils V_A at transmission distances of 50 mm, 100 mm, and 150 mm. The EMF leakage of the WPT system is measured to ensure that its value is lower than the ICNIRP standard. The corresponding V_A at different transmission distances is shown in Table 3.

Table 3. V_A at different transmission distances.

Transmission distance	V_A
50 mm	41.61 V
100 mm	24.32 V
150 mm	20.06 V

When the transmission distance is shortened, the magnetic flux density increases accordingly. Therefore, to meet the shielding requirements, the proposed dynamic shielding scheme is used to increase the V_A to 41.61 V and to shield from excessive EMF leakage. When the distance increases, the magnetic flux density becomes smaller correspondingly. At this point, based on the consideration of saving resources and economic costs, V_A is reduced to 20.06 V consequently, so that the EMF leakage is exactly within the safety limit of the ICNIRP standard, avoiding the unnecessary waste of resources.

The magnetic flux density distribution of the WPT system at different distances is displayed in in Figures 10 and 11, demonstrating the EMF leakage with no shielding and double-coil active shielding. The comparison indicates that the magnetic flux density increases with the shortening of the transmission distance and decreases with the increase in transmission distance, when no shielding is available. Moreover, the EMF leakage outside the system is higher. However, with the addition of double-coil active shielding, the EMF leakage outside the WPT system is significantly reduced by adjusting V_A, as illustrated in Figure 11.

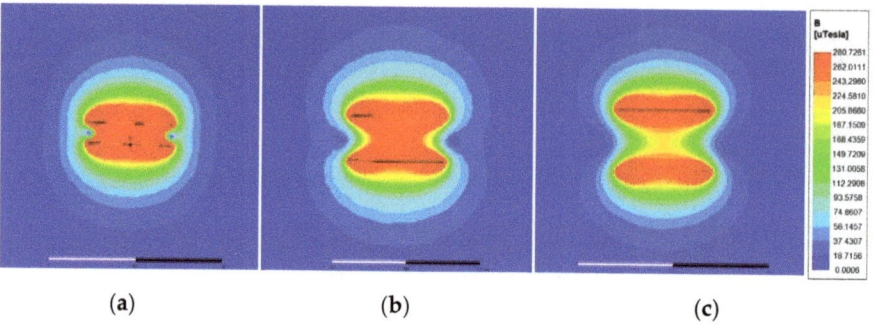

Figure 10. The magnetic flux density distribution of WPT system with no shielding: (**a**) 50 mm; (**b**) 100 mm; (**c**) 150 mm.

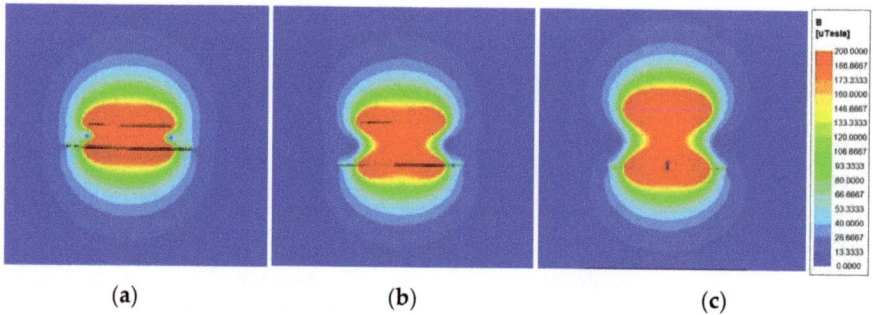

Figure 11. The magnetic flux density distribution of WPT system with double-coil active shielding: (**a**) 50 mm; (**b**) 100 mm; (**c**) 150 mm.

To further prove the shielding effectiveness of the dynamic shielding scheme and the impact of double-coil active shielding on the power transfer efficiency of the WPT system, the shielding effectiveness SE and power transfer efficiency η at different transmission distances after adjusting V_A are presented in Table 4.

Table 4. SE and η at different transmission distances.

Transmission distance	50 mm	100 mm	150 mm
SE	69.4%	76.2%	77.4%
η (no shielding)	96.5%	93.8%	92.7%
η (double-coil active shielding)	92.4%	91.9%	90.7%

It can be seen that the SE of this dynamic shielding scheme reaches more than 69.4%. Moreover, with the increase in distance, the shielding effectiveness becomes more prominent. At 150 mm, it achieves 77.4% shielding effectiveness for EMF leakage in the WPT system. Furthermore, by comparing the η of the WPT system with double-coil active shielding and no shielding, it can be found that the addition of double-coil active shielding employed in the paper allows the power transfer efficiency to be essentially above 90%. Therefore, the involvement of this double-coil active shielding structure has not caused a significant impact on the transmission efficiency, and the degradation of the performance of the WPT system is mitigated. In summary, the effectiveness of the proposed dynamic shielding scheme is successfully verified.

4.2. Experimental Verification

In order to validate the correctness of the simulation results and the effectiveness and feasibility of the proposed dynamic shielding scheme, the circuit of the WPT system with double-coil active shielding is built in this paper. The structural framework is shown in Figure 12, and a physical diagram of the experimental setup is displayed in Figure 13. And the general specifications of measuring devices are listed in Appendix A. Two control experimental setups are used to better visualize the characteristics of this scheme. One group involves no shielding, and the other group involves double-coil active shielding but without the dynamic shielding scheme.

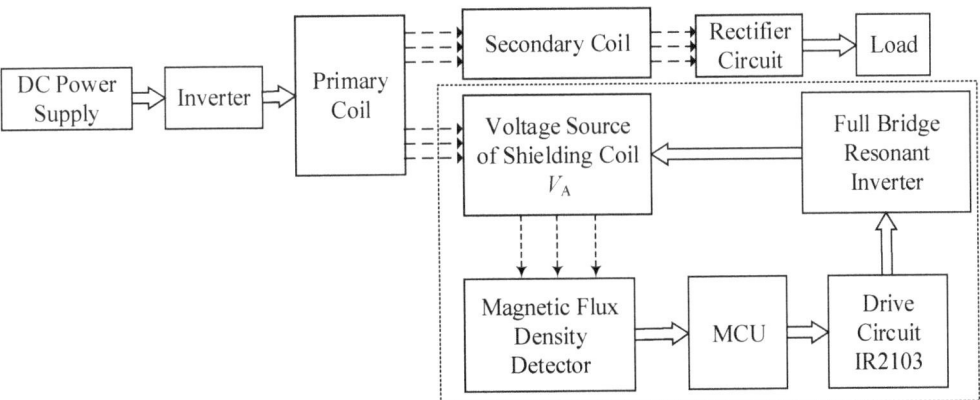

Figure 12. The structural framework of WPT system with double-coil active shielding.

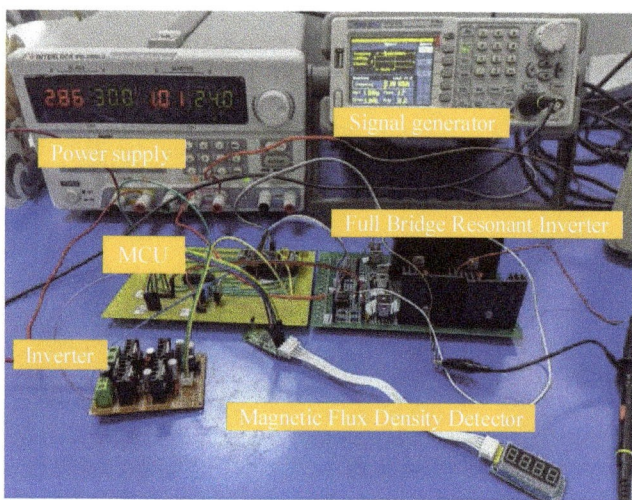

Figure 13. Experimental setup.

It mainly consists of four parts: (1) a power control module composed of DC power supply and an inverter; (2) a power transmission module composed of three coils—a transmitting coil, half-loop shielding coils and receiving coil; (3) a dynamic shielding control module composed of a full-bridge inverter circuit, driving circuit, magnetic flux density detector, and micro-controller chip (MCU); (4) a load module composed of a rectifier circuit and load. The power supply for the WPT system is provided by a DC power supply, which, through the inverter, forms a high-frequency AC signal to the transmitting coil. The power is transmitted from the transmitting coil to the active shielding coils and receiving coil and then finally rectified and converted to provide power for the load.

This work focuses on the dynamic shielding control module consisting of a full-bridge inverter circuit, driver circuit, magnetic flux density detector, and MCU (the dashed block part in the Figure 12). The magnetic flux density of EMF leakage is detected by the magnetic flux density detector, and it is output to the MCU; then, the MCU applies the dynamic shielding scheme to calculate the corresponding duty cycle and connects the required control signal to IR2103 to drive the conduction of four MOSFETs in the full-bridge inverter. Thus, the power supply V_A of the active shielding coils is controlled to realize the proposed dynamic shielding scheme.

Experiments are conducted by assigning different values to the transmission distance. The transmission distance is varied between 10 mm and 200 mm, and the variation step is set to 10 mm/step. The supply power V_A is adjusted depending on the change in transmission distance. In the following, experimental results of this dynamic shielding scheme are analyzed.

The curve of V_A variation with the transmission distance is shown in Figure 14. According to the ICNIRP standard, the magnetic flux density B is 27 μT. With the gradual increase in transmission distance, V_A decreases accordingly. In the range of 0–50 mm, V_A decreases sharply with the increase in transmission distance, and the decline is very apparent. It is related to the drastic magnetic flux density change in the near field of the WPT system. Within 50–150 mm, the decrease in V_A becomes slightly less, and within 150–200 mm, the change in V_A tends towards a stable value, which is related to the limited transmission distance of WPT technology.

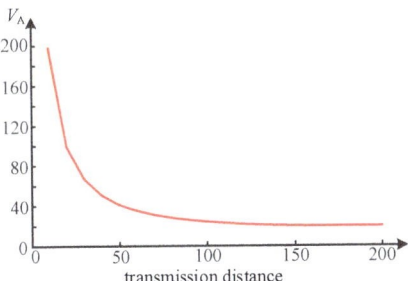

Figure 14. The V_A variation with transmission distance.

According to the guidance in Figure 14, V_A is adjusted so that the magnetic flux density reaches below 27 µT at different transmission distances. The variation curves of magnetic flux density at different transmission distances are given in Figure 15. Blue indicates no shielding, red represents the case with double-coil active shielding but without the dynamic shielding scheme (here V_A = 24 V and the initial transmission distance is 100 mm), and yellow represents the case with double-coil active shielding and with the dynamic shielding scheme to adjust V_A. Lines indicate calculated values, and dots indicate measured values.

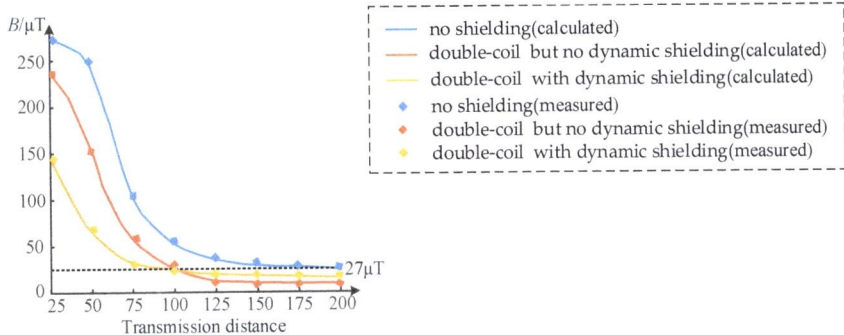

Figure 15. Curves of magnetic flux density at different transmission distances.

It can be seen from the three curves in Figure 15 that the measured and calculated values are in general agreement. Without shielding, the magnetic flux density within 10–163 mm is higher than the safety limit of 27 µT, and only in the range of 163–200 mm is it below the safety standard. When double-coil active shielding is used without the dynamic shielding scheme, the magnetic flux density is significantly higher than 27 µT at a distance less than the initial transmission distance of 100 mm. The greater the distance, the greater the flux density. Obviously, at reduced distances, the initial supply power V_A is not sufficient to shield the EMF leakage of the WPT system. Moreover, at a distance greater than the initial transmission distance, the magnetic flux density becomes gradually lower than the safety limit, when no corresponding adjustment of V_A will also result in a waste of resources. When adopting double-coil active shielding and applying the dynamic shielding scheme to adjust V_A, the EMF leakage can be basically kept below the safety limit of 27 µT regardless of the increase in or shortening of the transmission distance. It avoids excessive power waste when the distance increases and solves the problem of excessive EMF exposure to human safety.

A comparison of the shielding effectiveness SE with and without the dynamic shielding scheme for different transmission distances is shown in Figure 16. Measured and calculated values remain largely consistent. Without the dynamic shielding scheme, SE becomes progressively larger with increasing distance. This means that the closer one approaches the coil, the less EMF leakage is shielded. The higher the level of EMF exposure, the greater the risk to humans. This is due to the fact that V_A is not regulated accordingly. When the distance decreases, the magnetic flux density increases, and the original V_A is no longer enough to shield a sufficient amount of EMF leakage. On the other hand, with the dynamic shielding scheme, SE increases when the transmission distance is smaller and decreases slightly when the transmission distance becomes longer. The closer to the transmitting coil, the higher the magnetic flux density of EMF leakage, when a higher shielding effectiveness SE is required to bring the EMF leakage level down to within the safe range. The results show that this method is consistent with the limits of EMF exposure requirements and validate the effectiveness of the dynamic shielding scheme.

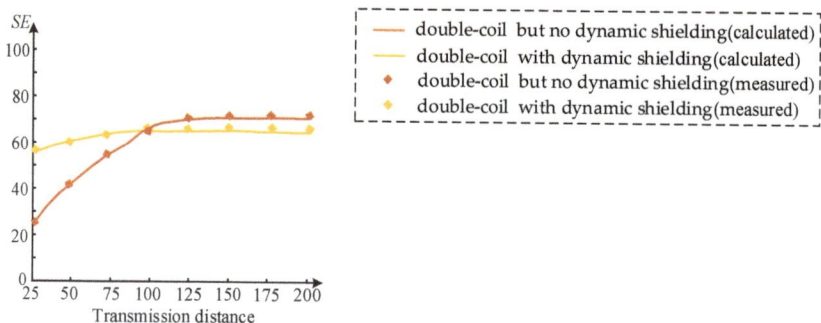

Figure 16. SE for different transmission distances with and without dynamic shielding scheme.

To further confirm the feasibility of this dynamic shielding scheme, the power transmission efficiency η of the WPT system is investigated. Figure 17 shows the variation curves of η at different transmission distances, where blue denotes no shielding and yellow denotes double-coil active shielding. It is clear that, although slightly fluctuating, the measured values match well with the calculated values. The η with double-coil active shielding is slightly reduced compared with that with no shielding, but the reduction is not significant and is approximately 3.1%. It is related to the introduction of the shielding system. It can be found that the addition of this double-coil active shielding structure achieves good shielding effectiveness against EMF leakage at the same time, without causing a significant sacrifice in power transmission efficiency. This means that the application of the double-coil dynamic shielding scheme can not only avoid the waste of power but also reduce the degrading influence of the shielding device on the transmission performance of the WPT system; thus, its feasibility is verified.

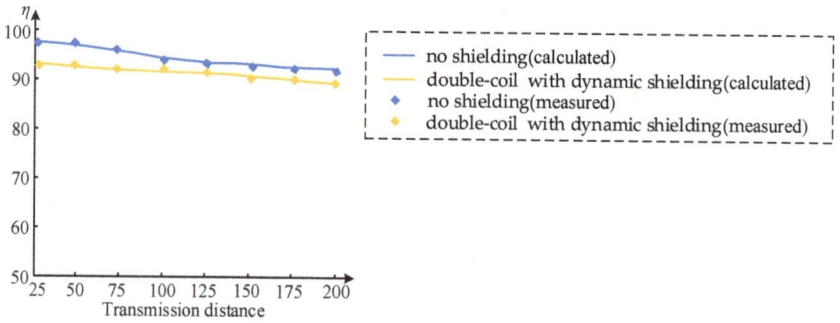

Figure 17. Curves of η at different transmission distances.

5. Conclusions

Given that different transmission distances lead to different degrees of EMF leakage, a double-coil dynamic shielding scheme based on active shielding technology is proposed in this paper. The active shielding coils adopt two half-loop structures and are set on the outside of the transmitting coil. The supply power of the active shielding coils installed on the ground is adjusted according to different transmission distances to achieve dynamic shielding of EMF leakage with different EVs. The WPT system with the proposed double-coil dynamic shielding scheme is modeled, simulated, experimented, and compared with other WPT systems.

The simulation and experimental results show that the proposed double-coil dynamic shielding scheme can shield approximately 77.4% of the EMF leakage and maintain high shielding effectiveness as the transmission distance varies. Furthermore, the application of the double-coil dynamic shielding scheme essentially has no effect on the power transmission efficiency. Therefore, the adaptability, effectiveness, and feasibility of the double-coil dynamic shielding scheme for WPT systems with different transmission distances are verified.

EVs currently available in the market have different structures, and the distances between their vehicle chassis and the ground are bound to be different. When the transmitting coil is fixed to the ground in the WPT system, EVs with different chassis heights signify different transmission distances. The proposed double-coil dynamic shielding scheme can shield the EMF leakage as the transmission distance changes, avoiding the repeated design of the shielding system. The proposed scheme is also generally applicable to other cases where the position of the transmitting coil is fixed but the distance of the receiving coil changes. Thus far, the scheme involves a large amount of analytical calculations, and it is hoped that a more concise method can be sought in future research.

Author Contributions: Conceptualization, Y.L. and Z.C.; methodology, Y.L.; software, Y.L.; validation, Y.L.; formal analysis, Y.L.; investigation, Y.L.; resources, Y.L.; data curation, Y.L.; writing—original draft preparation, Y.L.; writing—review and editing, Y.L.; visualization, Y.L.; supervision, S.Z.; project administration, Z.C.; funding acquisition, Z.C. and S.Z. All authors have read and agreed to the published version of the manuscript.

Funding: This research was funded by National Natural Science Foundation of China, grant number 61903272 and 61873180.

Institutional Review Board Statement: Not applicable.

Informed Consent Statement: Not applicable.

Conflicts of Interest: The authors declare no conflict of interest.

Nomenclature

WPT	wireless power transfer
EV	electric vehicle
EMF	electromagnetic field
ICNIRP	International Commission on Non-Ionizing Radiation Protection
r_{in}	inner radius of half-loop
r_{out}	outer radius of half-loop
r_s	radius of transmitting coil
B_M	total magnetic flux density at point M
d_{S1}	distance from the transmitting coil to the point M
d_{S2}	distance from the receiving coil to point M
B_1	magnetic flux density of transmitting coil
B_2	magnetic flux density of receiving coil
B_{3_in}	magnetic flux density of the internal shielding coils at point M
B_{3_out}	magnetic flux density of the external shielding coils at point M
SE	shielding effectiveness

$R_{1,2,3}$	internal resistances of the transmitting, receiving, and shielding coils
$L_{1,2,3}$	self-inductances of the transmitting, receiving, and shielding coils
$C_{1,2,3}$	capacitances of the transmitting, receiving, and shielding coils
R_L	resistive load
R_S	internal resistance of power supply
M_{12}	mutual inductances between transmitting and receiving coils
M_{13}	mutual inductances between transmitting and shielding coils
M_{23}	mutual inductances between receiving and shielding coils
$I_{1,2,3}$	current of transmitting, receiving, and shielding coils
ω	angular operating frequency
V_S	power supply of transmitting coil
V_A	power supply of shielding coil
$P_{1,2,3}$	output power of transmitting, receiving, and active shielding coils
PWM	pulse width modulation
α	duty cycle
N	number of turns
MCU	micro-controller chip
η	power transfer efficiency

Appendix A

Table A1. General specifications of measuring devices used in the WPT experiment.

Devices	Product Model	Operating Parameters
Power Supply	IPD-3303LU	0~32 V
Magnetic Flux Density Detector	SS49E	−1000~1000 Gs
Signal Generator	SDG830	1 μHz~30 MHz
Drive Circuit	IR2103	10~20 V
MCU	STM32F103C8T6	2~3.6V
MOSFET	IRF640	−20~20 V

Appendix B

Table A2. Features of various shielding technologies in WPT.

Shielding Technology	Advantages	Disadvantages
Passive Shielding	simple operation, easy implementation	large volume, high space occupation, increased loss
Resonant Reactive Current Loop	small additional volume, simple structure	lower shielding effectiveness
Active Shielding	high shielding effectiveness, small additional volume	complicated calculations

References

1. Triviño, A.; Gonzalez-Gonzalez, J.; Castilla, M. Review on Control Techniques for EV Bidirectional Wireless Chargers. *Electronics* **2021**, *10*, 1905. [CrossRef]
2. Arif, S.; Lie, T.; Seet, B.; Ayyadi, S.; Jensen, K. Review of Electric Vehicle Technologies, Charging Methods, Standards and Optimization Techniques. *Electronics* **2021**, *10*, 1910. [CrossRef]
3. Sanguesa, J.; Torres-Sanz, V.; Garrido, P.; Martinez, F.; Marquez-Barja, J. A Review on Electric Vehicles: Technologies and Challenges. *Smart Cities* **2021**, *4*, 372–404. [CrossRef]
4. Li, S.; Mi, C.C. Wireless power transfer for electric vehicle applications. *IEEE J. Emerg. Sel. Top. Power Electron.* **2014**, *3*, 4–17.
5. Bertoluzzo, M.; Di Monaco, M.; Buja, G.; Tomasso, G.; Genovese, A. Comprehensive Development of Dynamic Wireless Power Transfer System for Electric Vehicle. *Electronics* **2020**, *9*, 1045. [CrossRef]
6. Colussi, J.; La Ganga, A.; Re, R.; Guglielmi, P.; Armando, E. 100 kW Three-Phase Wireless Charger for EV: Experimental Val-idation Adopting Opposition Method. *Energies* **2021**, *14*, 2113. [CrossRef]
7. Chen, K.; Cheng, K.; Yang, Y.; Pan, J. Stability Improvement of Dynamic EV Wireless Charging System with Receiver-Side Control Considering Coupling Disturbance. *Electronics* **2021**, *10*, 1639. [CrossRef]

8. Zhu, Q.; Zhang, Y.; Guo, Y.; Liao, C.; Wang, L.; Wang, L. Null-Coupled Electromagnetic Field Cancelling Coil for Wireless Power Transfer System. *IEEE Trans. Transp. Electrif.* **2016**, *3*, 464–473. [CrossRef]
9. Mohamed, A.; Marim, A.A.; Mohammed, O. Magnetic Design Considerations of Bidirectional Inductive Wireless Power Transfer System for EV Applications. *IEEE Trans. Magn.* **2017**, *53*, 1–5. [CrossRef]
10. Do, C.Y.; Park, E.Y. Impact investigations and characteristics by strong electromagnetic field of wireless power charging system for electric vehicle under air and water exposure indexes. *IEEE Trans. Appl. Supercond.* **2018**, *28*, 1–5.
11. Yamazaki, K.; Taki, M.; Ohkubo, C. Safety assessment of human exposure to intermediate frequency electromagnetic fields. *Electr. Eng. Jap.* **2016**, *197*, 3–11. [CrossRef]
12. Campi, T.; Cruciani, S.; De Santis,, V. EMC and EMF safety issues in wireless charging system for an electric vehicle (EV). In Proceedings of the 2017 International Conference of Electrical and Electronic Technologies for Automotive, Turin, Italy, 27 July 2017; IEEE: Piscataway, NJ, USA, 2017; pp. 1–4.
13. Miyakoshi, J.; Tonomura, H.; Koyama, S.; Narita, E.; Shinohara, N. Effects of Exposure to 5.8 GHz Electromagnetic Field on Micronucleus Formation, DNA Strand Breaks, and Heat Shock Protein Expressions in Cells Derived from Human Eye. *IEEE Trans. NanoBiosci.* **2019**, *18*, 257–260. [CrossRef]
14. Asa, E.; Mohammad, M.; Onar, O.C.; Pries, J.; Galigekere, V.; Su, G.-J. Review of Safety and Exposure Limits of Electromagnetic Fields (EMF) in Wireless Electric Vehicle Charging (WEVC) Applications. In Proceedings of the 2020 IEEE Transportation Electrification Conference & Expo (ITEC), Chicago, IL, USA, 23–26 June 2020; pp. 17–24. [CrossRef]
15. Shin, Y.; Park, J.; Kim, H.; Woo, S.; Park, B.; Huh, S.; Lee, C.; Ahn, S. Design Considerations for Adding Series Inductors to Reduce Electromagnetic Field Interference in an Over-Coupled WPT System. *Energies* **2021**, *14*, 2791. [CrossRef]
16. Tang, L.-C.; Jeng, S.-L.; Chang, E.-Y.; Chieng, W.-H. Variable-Frequency Pulse Width Modulation Circuits for Resonant Wireless Power Transfer. *Energies* **2021**, *14*, 3656. [CrossRef]
17. Feliziani, M.; Cruciani, S.; Campi, T.; Maradei, F. Near Field Shielding of a Wireless Power Transfer (WPT) Current Coil. *Prog. Electromagn. Res. C* **2017**, *77*, 39–48. [CrossRef]
18. Lee, S.; Jeong, S.; Hong, S.; Sim, B.; Kim, J. Design and Analysis of EMI Shielding Method using Intermediate Coil for Train WPT System. In Proceedings of the 2018 IEEE Wireless Power Transfer Conference (WPTC), Montreal, QC, Canada, 3–7 June 2018; pp. 1–4. [CrossRef]
19. Zhang, B.; Carlson, R.B.; Galigekere, V.P.; Onar, O.C.; Pries, J.L. Electromagnetic Shielding Design for 200 kW Stationary Wireless Charging of Light-Duty EV. In Proceedings of the IEEE Energy Conversion Congress and Exposition (ECCE), Detroit, MI, USA, 1–15 October 2020; pp. 5185–5192. [CrossRef]
20. Kim, J.; Kim, J.; Kong, S. Coil design and shielding methods for a magnetic resonant wireless power transfer system. *Proc. IEEE* **2013**, *101*, 1332–1342. [CrossRef]
21. Tan, L.; Elnail, K.E.I.; Ju, M.; Huang, X. Comparative Analysis and Design of the Shielding Techniques in WPT Systems for Charging EVs. *Energies* **2019**, *12*, 2115. [CrossRef]
22. Mohammad, M.; Wodajo, E.T.; Choi, S.; Elbuluk, M.E. Modeling and Design of Passive Shield to Limit EMF Emission and to Minimize Shield Loss in Unipolar Wireless Charging System for EV. *IEEE Trans. Power Electron.* **2019**, *34*, 12235–12245. [CrossRef]
23. Wen, F.; Huang, X. Optimal Magnetic Field Shielding Method by Metallic Sheets in Wireless Power Transfer System. *Energies* **2016**, *9*, 733. [CrossRef]
24. Li, J.; Huang, X.; Chen, C.; Tan, L.; Wang, W.; Guo, J. Effect of metal shielding on a wireless power transfer system. *AIP Adv.* **2017**, *7*, 056675. [CrossRef]
25. Kim, S.; Park, H.-H.; Kim, J.; Kim, J.; Ahn, S. Design and Analysis of a Resonant Reactive Shield for a Wireless Power Electric Vehicle. *IEEE Trans. Microw. Theory Tech.* **2014**, *62*, 1057–1066. [CrossRef]
26. Park, J.; Kim, D.; Hwang, K. A resonant reactive shielding for planar wireless power transfer system in smartphone applica-tion. *IEEE Trans. Electromagn. Compat.* **2017**, *59*, 695–703. [CrossRef]
27. Nie, Y.; Jiao, C.; Fan, Y. Active Shielding Design of Patrol Robot Wireless Charging System. In Proceedings of the 2019 IEEE 3rd International Electrical and Energy Conference (CIEEC), Beijing, China, 7–9 September 2019; IEEE: Piscataway, NJ, USA, 2019; pp. 2003–2007.
28. Campi, T.; Cruciani, S.; Maradei, F.; Feliziani, M. Active Coil System for Magnetic Field Reduction in an Automotive Wireless Power Transfer System. In Proceedings of the 2019 IEEE International Symposium on Electromagnetic Compatibility, Signal & Power Integrity (EMC + SIPI), New Orleans, LA, USA, 22–26 July 2019; pp. 189–192. [CrossRef]
29. Kim, J.; Ahn, J.; Huh, S.; Kim, K.; Ahn, S. A Coil Design and Control Method of Independent Active Shielding System for Leakage Magnetic Field Reduction of Wireless UAV Charger. *IEICE Trans. Commun.* **2020**, *103*, 889–898. [CrossRef]
30. Choi, S.Y.; Gu, B.W.; Lee, S.W.; Lee, W.Y.; Huh, J.; Rim, C.T. Generalized Active EMF Cancel Methods for Wireless Electric Vehicles. *IEEE Trans. Power Electron.* **2013**, *29*, 5770–5783. [CrossRef]
31. Triviño, A.; González-González, J.; Aguado, J. Wireless Power Transfer Technologies Applied to Electric Vehicles: A Review. *Energies* **2021**, *14*, 1547. [CrossRef]
32. De Santis, V.; Giaccone, L.; Freschi, F. Chassis Influence on the Exposure Assessment of a Compact EV during WPT Recharging Operations. *Magnetochemistry* **2021**, *7*, 25. [CrossRef]
33. Kim, D.-H.; Kim, M.-S.; Kim, H.-J. Frequency-Tracking Algorithm Based on SOGI-FLL for Wireless Power Transfer System to Operate ZPA Region. *Electronics* **2020**, *9*, 1303. [CrossRef]

34. Aldhaher, S.; Luk, P.; Whidborne, J. Tuning Class E Inverters Applied in Inductive Links Using Saturable Reactors. *IEEE Trans. Power Electron.* **2013**, *29*, 2969–2978. [CrossRef]
35. Uddin, M.K.; Ramasamy, G.; Mekhilef, S.; Ramar, K.; Lau, Y.-C. A review on high frequency resonant inverter technologies for wireless power transfer using magnetic resonance coupling. In Proceedings of the 2014 IEEE Conference on Energy Conversion (CENCON), Johor Bahru, Malaysia, 13–14 October 2014; pp. 412–417. [CrossRef]
36. Asl, E.S.; Babaei, E.; Sabahi, M.; Nozadian, M.H.B.; Cecati, C. New Half-Bridge and Full-Bridge Topologies for a Switched-Boost Inverter with Continuous Input Current. *IEEE Trans. Ind. Electron.* **2017**, *65*, 3188–3197. [CrossRef]
37. Vassilev, A.; Ferber, A.; Wehrmann, C.; Pinaud, O.; Schilling, M.; Ruddle, A.R. Magnetic Field Exposure Assessment in Electric Vehicles. *IEEE Trans. Electromagn. Compat.* **2014**, *57*, 35–43. [CrossRef]

Article

Reduction of Cogging Torque in Surface Mounted Permanent Magnet Brushless DC Motor by Adapting Rotor Magnetic Displacement

T. A. Anuja and M. Arun Noyal Doss *

Department of Electrical and Electronics Engineering, SRM Institute of Science and Technology, SRM Nagar, Kattankulathur 603203, India; taanuja@gmail.com
* Correspondence: arunnoyal@gmail.com or arunoyad@srmist.edu.in

Abstract: Cogging torque is a critical dilemma in Permanent Magnet Brushless DC (PMBLDC) motors. In medium-low power PMBLDC motors, redundant vibrations and forbidding noises arise as a result of the harmonic magnetic forces created by cogging torque. This paper introduces a simple approach for minimizing cogging torque in PMBLDC motors by applying placement irregularities in rotor magnets. An angle shift in the rotor magnets in surface-mounted PMBLDC motors helps to attain magnet displacement. This displacement imparts an asymmetrical magnet structure to the rotor. Maintaining pole arc to pole pitch ratio (L/τ) of between 0.6 and 0.8, shifting angles from $1°$ to $8°$ were considered in order to analyze the effect of the angle shift on the rotor magnets. An analytical expression was also derived for finding the shifting angle with the minimum cogging torque in the PMBLDC motor by using the Virtual Work Method (VWM). The optimization of the shifting angle with minimum cogging torque was investigated using 3D Finite Element Analysis (FEA). A comparison of the simulation and analytical results of cogging torque was carried out. It was determined that the reduction of cogging torque in the analytical results showed good agreement with the FEA analysis.

Keywords: PMBLDC motor; cogging torque; finite element analysis; virtual work method; shifting angle

1. Introduction

Permanent Magnet Brushless DC (PMBLDC) motors are machines with excellent torque–speed characteristics, excellent efficiency, and nominal maintenance cost. They are very favorable for unidirectional variable-speed applications such as automotive pumps and fans [1]. They are also very supportive with respect to achieving compactness in terms of machine size [2]. They produce a negligible amount of electromagnetic and mechanical noise. They also exhibit excellent durability due to their lack of mechanical contact [3]. One of the great flaws in the BLDC motor is its torque ripple, which is built into the design [4]. One of the main reasons for torque ripple is cogging torque. Cogging torque also generates enormous, troubling noise and shaking movements in the machine itself, as well as in its load. For this reason, cogging torque reduction methods play a vital role in PMBLDC motor design. Generally, cogging torque derives from motors using permanent magnets such as Permanent Magnet Brushless DC (PMBLDC) motors and permanent Magnet Synchronous motors. The root cause for the generation of cogging torque in BLDC motors is the magnetic interaction between the permanent magnet and the steel in the slotted armature. A lot of techniques are currently available for the minimization of cogging torque in BLDC motors. Nowadays, BLDC motors are used in rigorous applications such as electric power steering, robotics, etc. Hence the reduction of cogging torque has come to be a grueling task.

Over the last two decades, numerous studies have been conducted on the cogging torque of PMBLDC motors. These methods have included electromagnetic methods and mechanical methods. The authors recommended a core skew structure for reducing the cogging torque in [5]. When applying this property, no-load THD and the fifth and seventh harmonics were the main harmonic components of the back-EMF. However, when skewing the core, some residual magnetism is lost. Modification of the machine's magnetic circuit resulted in the occurrence of additional harmonics in the cogging torque in [6]. They focused on a symmetrical rotor with an asymmetrical stator or an asymmetrical rotor with a symmetrical stator. Nonlinear algorithms were proposed for the reduction of cogging torque in [7–9]. A feedback linearization strategy was used in [7], mathematical modeling of cogging torque phenomena was performed in [8], and field-oriented control operations were used in [9]. A new technique for radial flux surface-mounted PMBLDC motors was proposed by applying T-shaped bifurcations in the stator teeth in [10]. However, this reduced the mechanical strength of the stator. In order to reduce the cogging torque, the slot was closed by a sliding separator in [11]. For small-sized machines, it is very difficult to place a slider inside the slot. The authors recommended notches in the rotor in [12]. However, adding notches in the rotor side is a tedious mechanical process. A new technique in the winding side was proposed in [13]. Coil winding concentrated on the phase group is a good solution for cogging torque reduction, but increases the complexity of the winding. Magnet step skewing and reduction of the claw pole width was carried out in a claw pole machine in [14]. Magnet step skewing causes a reduction in residual magnetism and the unequal width in the claw pole increases the structural complexity. U-clamped magnetic poles were recommended for the reduction of cogging torque and even flux per pole in [15], so a machine with a greater number of slots is recommended, as well as a high magnet thickness. In [16], a novel air gap profile was introduced for a single-phase PMBLDC motor. This air gap profile consists of a dip and a dip angle. By varying the dip and dip angle, a handful of air gap profiles could be generated, and profiles with a dip angle less than the critical dip angle exhibited improved starting torque by up to 70%. In [17], the reduction of cogging torque and acoustic noise in permanent magnet motors with larger stator slot openings was investigated. Here also, tooth pairings with two different types of tooth width were proposed. The experimental results showed that the proposed tooth pairings reduced cogging torque by 85% and acoustic noise by 3.1 dB. Reference [1] showed how to minimize high cogging torque without increasing the manufacturing cost. A systematic means was presented by which the selected introduction of auxiliary slots can double the fundamental frequency of the cogging torque, making stator claw skewing much more effective at reducing cogging torque; both measures can feasibly be carried out at the stage of punching the steel sheets and their subsequent deep drawing, at no additional cost. Reference [18] reported a stator shape optimization design for reducing the cogging torque of single-phase brushless DC (BLDC) motors by adopting an asymmetrical airgap to make them self-start. A model that combined Latin hypercube sampling and a genetic algorithm was used to reduce the cogging torque and maintain the efficiency and torque. As an optimal design result, the cogging torque of the optimal model decreased. Ref. [19] proposed a new design of SPOKE-type PM brushless direct current (BLDC) motor without using neodymium PM (Nd-PM). The proposed model had an improved output characteristic, as it used the properties of the magnetic flux effect of the SPOKE-type motor with an additional pushing assistant magnet and sub assistant magnet in the shape of a spoke. In this paper, ferrite PM (Fe-PM) was used instead of Nd-PM. The authors of [20] introduced a genetic algorithm for optimal core shape design for reducing cogging torque in brushless DC motors used in digital versatile disk drive systems or hard disk drive systems. The optimized or rounded core could be a recommended core shape for the outer-rotor-type BLDC motor for a DVD ROM drive system in order to achieve low cogging torque. Ref. [21] described a novel rotor pole shape consisting of a uniform surface and an eccentric surface, leading to a sinusoidal magnetic flux density in the air gap and reducing cogging torque, torque ripple, and the harmonics of the back-electromotive force waveform in a spoke-type brushless DC motor. This novel rotor included an eccentric surface. The proposed method had a smooth variation of reluctance, producing a near sinusoidal magnetic flux waveform in the air gap. This caused a reduction in the losses between the upper edge of the permanent magnet and the adjacent stator teeth, and an increment of the effective flux by concentrating the flux. Ref. [22] proposed an anisotropic ferrite magnet shape and magnetization direction to maximize back-EMF in an IPM BLDC motor. Firstly, four different models of general magnet shapes were selected, and then FEM analysis was carried out using four different magnetization directions for each of the four models. The best magnet shape and magnetization direction for each model was used to determine an initial model for optimization. Secondly, based on the initial model, optimization design for maximum back-EMF and minimum cogging torque and THD was performed. Reference [23] presented the stator and rotor shape designs for an interior-permanent magnet (IPM)-type brushless DC (BLDC) motor for reducing torque fluctuation. A partly enlarged air gap is introduced by the unequal diameter of the rotor and core structure of the stator with pole shoe modification. A reduction in torque ripple was obtained by upgrading the torque value at the minimum torque position, and their detail characteristics were compared. The addition of holes in the rotor core is a better solution for overcoming this problem. The additional torque fluctuation was decreased.

In the context of the exploration of different cogging torque reduction methods, motors with asymmetrical magnets represent a productive technique. This paper introduces a novel design based on an asymmetrical rotor structure by applying a shifting angle to the PM. The shifting angle method is not a skewing method, where the skew technique is used for skewing the rotor PMs. The shifting angle method alters the position of the permanent magnet by changing the pole pitch. To analyze the effect of magnets shifting in the BLDC motor, 3D-FEA analysis and numerical analysis were performed. To study the impact of varying the angle between the permanent magnet on the rotor, a range of different angles were considered. The rotor permanent magnet was shifted from 1° to 8°. To effectively decrease the cogging torque, an optimal 3° shift to the permanent magnet (AB-87°, BC-87°, CD-93°, and DA-93°) was determined. The resulting design also obtained the correct trapezoidal shape of the back-EMF. Compared to the reference model, it was able to achieve a 60% reduction in cogging torque. This method is able to achieve excellent performance characteristics. The simulation results were compared with the analytical results, and the reduction of cogging torque in the two analyses was almost the same.

The organization of the paper is as follows: the design of the BLDC motor is presented in Section 2. The analytical expression of cogging torque by VWM is presented in Section 3. Finite element analysis of the symmetrical rotor is presented in Section 4. The effect of the angle shift on cogging torque is shown in Section 5.

2. Design of BLDC Motor

The basic step in the motor design is to fix the rated speed and torque. Based on these particulars, the other parameters of the motor can be selected. Figure 1 shows the structure of a surface-mounted PMBLDC motor.

The stators of BLDC motors are similar to those of three-phase induction motors. The main dimensions depend on the specific electric loading and specific magnetic loading. The procedure for calculating the main dimensions is depicted as a flowchart in Figure 2. The procedure for the selection of stator slots and winding is shown in Figure 3.

In this paper, in order to analyze cogging torque reduction, a 4-pole, 12-slot surface-mounted BLDC motor was used. The design parameters of the motor are shown in Table 1.

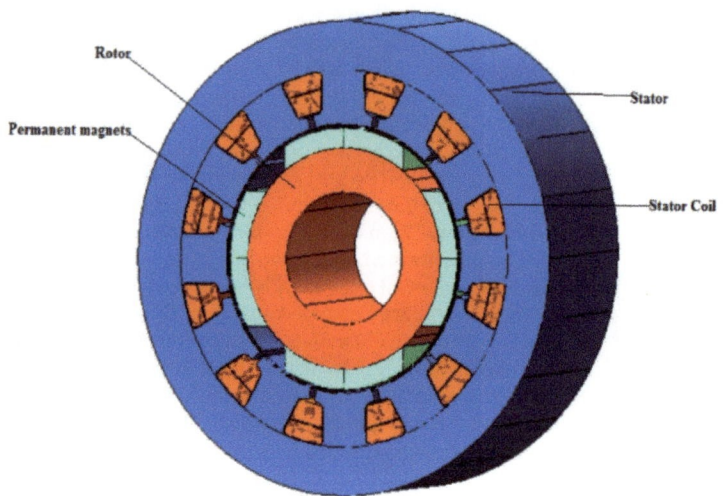

Figure 1. Topology of surface-mounted BLDC motor.

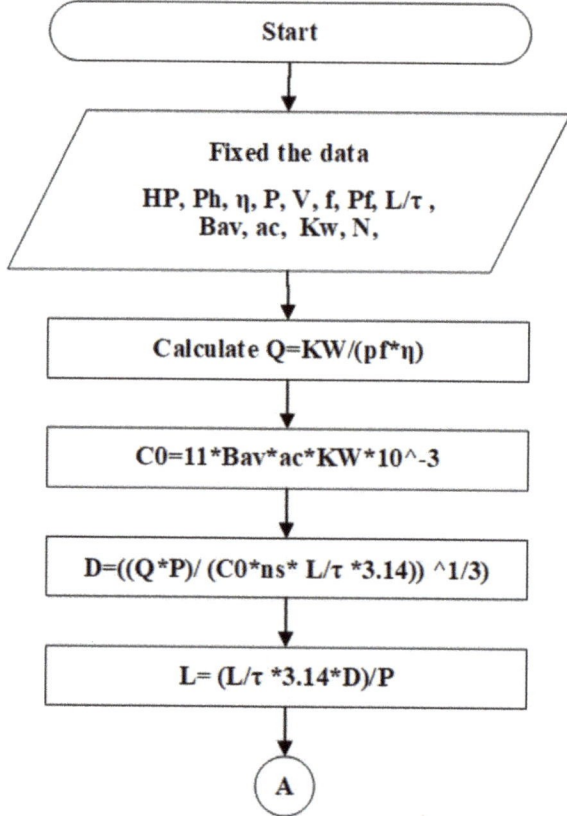

Figure 2. Flowchart for calculation of main stator dimensions.

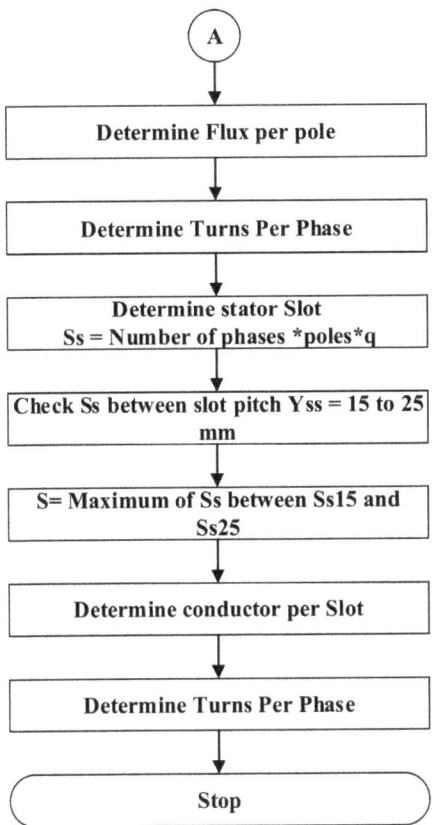

Figure 3. Flowchart for stator slots and windings.

Table 1. Design parameters.

Parameters	Rating	Parameters	Rating
Supply Voltage (v)	48	Power (HP)	1
Number of poles	4	Number of Slots	12
Rated Speed (rpm)	5000	Inner Diameter Rotor (mm)	16.7
Outer Diameter (mm)	60	Outer Diameter Rotor (mm)	33
Stack Height (mm)	50	Airgap Thickness (mm)	0.5
Pole Arc	63°	Magnet Thickness (mm)	2.5

3. Analytical Expression of Cogging Torque by VWM

The most widely used method for calculating cogging torque is the virtual work method (VWM). The virtual work method is also called the co-energy method [24]. The basic principle is that neglecting the variation in magnets and iron of PMBLDC motor, the cogging torque of an ideal lossless PMBLDC motor can be expressed as the derivative of co-energy in the air gap [25]. According to [26], cogging torque is produced because of the interaction between the PMs and the armature and the slot. Because of this interaction in the absence of a current, energy variation occurs inside the motor.

$$E_v = E_{v.I} + E_{v.airgap} + E_{v.PM} \tag{1}$$

where E_v is the total energy variation, $E_{v.I}$ is the energy variation in iron, $E_{v.airgap}$ is the energy variation in airgap, and $E_{v.PM}$ is the energy variation in PM.

When compared with the energy variation in the airgap and PM, only a minor variation occurs in iron. Therefore

$$E_v \cong E_{v.airgap} + E_{v.PM} = \frac{1}{2\mu_0} \iiint B^2 dv \tag{2}$$

Hence, cogging torque can be stated as

$$T_{cog} = -\frac{\partial E_v}{\partial \alpha} \tag{3}$$

where, μ_0, B and α are the permeability of air, the magnetic flux density (magnetic induction) and the angle of rotation of the rotor, respectively. The distribution of magnetic induction certainly stated as

$$B(\theta, \alpha) = B_{rs}(\theta) \frac{l_m}{l_m + l_g(\theta, \alpha)} \tag{4}$$

where $B_{rs}(\theta)$ is the residual flux density along the periphery of the airgap, l_m is the length of the permanent magnet, and l_g is the effective length of airgap distribution. Equation (2) can be redrafted as

$$E_v = \frac{1}{2\mu_0} \iiint B_{rs}^2(\theta) \left[\frac{l_m}{l_m + l_g(\theta, \alpha)}\right]^2 dv \tag{5}$$

To obtain the magnetostatic energy within the motor, Fourier expansion of $B_{rs}^2(\theta)$ and $\left[\frac{l_m}{l_m+l_g(\theta,\alpha)}\right]^2$ can be performed.

$$B_{rs}^2(\theta) = B_{rs0} + \sum_{n=1}^{\infty} B_{rsan} \cos n\theta + B_{rsbn} \sin n\theta \tag{6}$$

$$\left[\frac{l_m}{l_m + l_g(\theta, \alpha)}\right]^2 = G_0 + \sum_{n=1}^{\infty} G_n \cos ns(\theta + \alpha) \tag{7}$$

The analytical statement of cogging torque for asymmetrical magnets can be expressed as

$$T_{cog} = \frac{\pi z L s}{4\mu_0}\left(R_r^2 - R_s^2\right) \sum_{n=1}^{\infty} B_{rsanz} \sin ns\alpha + B_{rsbnz} \cos ns\alpha \tag{8}$$

where Ls is the length of the stack, s is the slot number, Rr is the rotor outer radius and Rs is the stator inner radius. The Fourier coefficients B_{rsanz} and B_{rsbnz} can be expressed as

$$B_{rsanz} = \frac{2B_{rs}^2}{ns\pi} \sin \frac{ns\pi\alpha_p}{2p} \sum_{k=1}^{2p} \cos ns \left[\frac{\pi}{p}(k-1) + \theta_s\right] \tag{9}$$

$$B_{rsbnz} = \frac{2B_{rs}^2}{ns\pi} \sin \frac{ns\pi\alpha_p}{2p} \sum_{k=1}^{2p} \sin ns \left[\frac{\pi}{p}(k-1) + \theta_s\right] \tag{10}$$

where, θ_s is the shifting angle.

When symmetrical structure is adopted for the magnets ($\theta_s = 0°$), B_{rsbnz} is zero and cogging torque can be expressed as

$$T_{cog} = \frac{\pi z Ls}{4\mu_0}\left(R_r^2 - R_s^2\right) \sum_{n=1}^{\infty} B_{rsanz} \sin ns\alpha \tag{11}$$

Table 2 shows the analytical results of cogging torque with shifting angles ranging from 1° to 8°. Figure 4 shows a graphical representation of cogging torque with shifting angles ranging from 1° to 8°.

Table 2. Analytical results for cogging torque with shifting angles from 1° to 8°.

Shifting Angle (°)	1	2	3	4	5	6	7	8
Cogging torque (Nm)	0.39	0.3	0.15	0.18	0.45	0.2	0.6	0.51

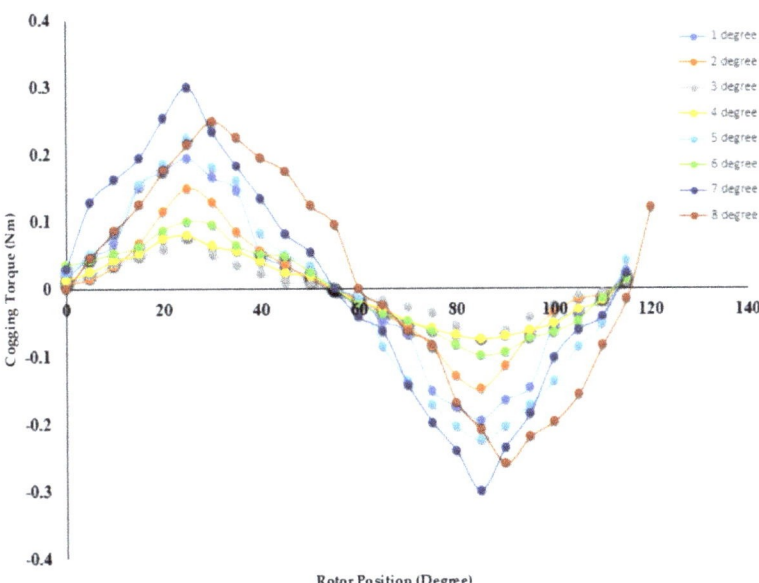

Figure 4. Analytical results of cogging torque with shifting angles from 1° to 8°.

4. Finite Element Analysis of a Symmetrical Rotor

A variety of numerical methods are available for the analysis of electromagnetic torque. These include, the virtual work method, the Maxwell stress tensor method, the nodal force method, and the Coulomb virtual work method, which are all methods used to analyze the cogging torque of PMBLDC motors. For the determination of cogging torque, precise field solutions are required. That is to say that a sophisticated mesh discretization is vital in FEA analysis, just as a dependable physical model is crucial for analytical calculations.

This section describes the outcomes of the 3D FEA analysis of the examined PMBLDC motor with the symmetrical rotor. Cogging torque can be determined numerically using the FEA method. Without a prototype, FEA is a powerful and viable tool for defining the performance of a given design [27,28]. The main target of this work was to decrease the cogging torque by using the magnet displacement method.

A 12-slot, 4-pole PMBLDC motor was considered as the motor for which the cogging torque was determined analytically using the FEA method. Further, on the basis of the FEA analysis, a comparative study was performed to compare the effect of the magnet shifting with the symmetrical structure. Figure 5 shows the conventional rotor with symmetrical magnets. Table 3 shows the FEA results for the symmetrical structure.

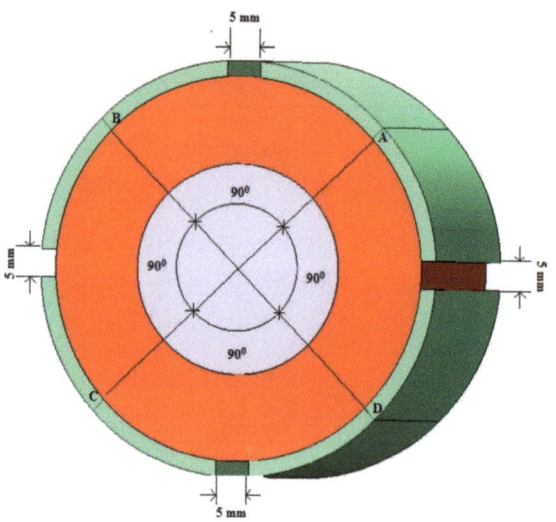

Figure 5. Conventional rotor with symmetrical magnets.

Table 3. FEA results for BLDC motor with symmetrical magnets.

Shifting Angle and Spacing between Permanent Magnets									Cogging Torque (Nm)
A-B		B-C		C-D		D-A			
Angle (°)	Space (mm)	Angle (°)	Space (mm)	Angle (°)	Space (mm)	Angle (°)	Space (mm)		
90	5	90	5	90	5	90	5		0.64

5. Effect of Angle Shift on Cogging Torque

In practice, magnet poles are neither made identically nor placed at the perfect location. There are a lot of methods available to obtain asymmetrical rotor structures. In this work, the method of shifting angle between the poles was adopted. Maintaining the L/τ ratio between 0.6 and 0.8, shifting angles were considered between 1° and 8°. From the analysis, it is understandable that the cogging torque is very sensitive to variations in the magnet angle. Figure 6 shows the procedure for finding the shifting angle with the minimum cogging torque.

In order to minimize cogging torque, a permanent magnet shift is considered here. A range of distinct angles were considered to analyze the effect of the angle shift on the rotor magnets. The permanent magnets on the rotor were shifted from 1° to 8°. The actual position of the permanent magnets was 90°. For every 1° change, four possible combinations were obtained. Table 4 shows the tabulation of the results of cogging torque with variation of magnet shifting from 1° to 8°. It shows that the shifting angle with 3° has the minimum cogging torque. The lowest cogging torque is 0.16 Nm when the angle between the poles A-B is 87°, B-C is 93°, C-D is 87° and D-A is 93°. Figure 7 shows the placement of the rotor magnet after applying the magnet shift to the rotor magnets. From the figure, it is clear that when the poles A, B, C, and D have a 90° difference with respect to each other, then each magnet has a 5 mm difference. Figure 7a shows shifting angle between the magnet is 1°. The spacing between A-B and C-D 4.44 mm and B-C and D-A is 5.56 mm. Figure 7b shows shifting angle between the magnet is 2°. The spacing between A-B and C-D is 5.58 mm and B-C and D-A is 4.42 mm. In Figure 7c the shifting angle between the magnet is 3°. The spacing between A-B and C-D is 3.32 mm and B-C and D-A is 6.68 mm. in Figure 7d the shifting angle between the magnet is 4°. The spacing between A-B and C-D is 7.24 mm and B-C and D-A is 2.76 mm. Figure 7e shows shifting angle between the magnet is 5°. The spacing between A-B and C-D is 2.2 mm and B-C and D-A is 7.8 mm. Figure 7f shows shifting angle between the magnet is 6°. The spacing between A-B and C-D is 8.38mm and B-C and D-A is 1.62 mm. In Figure 7g the shifting angle between the magnet is 7°. The spacing between A-B and C-D is 8.95 mm and B-C and D-A is 1.05 mm. When the shift angle becomes 8° in Figure 7h, the spacing between A-B and C-D is 9.52 mm, and the spacing between B-C and D-A is 0.48 mm. If we again increase the shift angle to 9°, the two magnets merge, and will act like a two-pole machine.

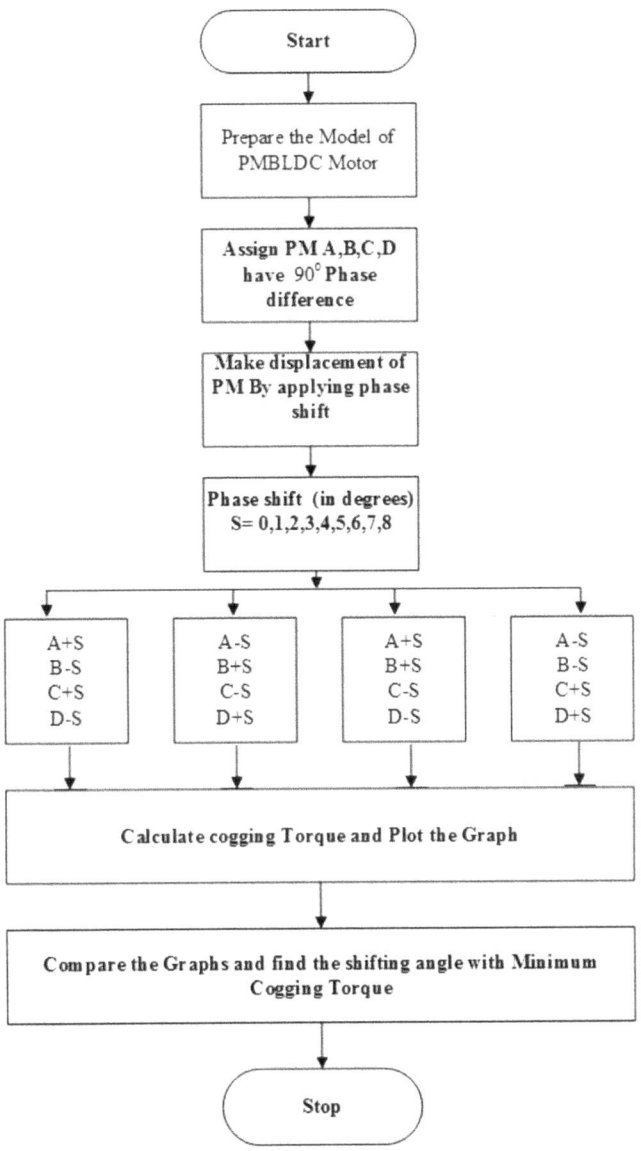

Figure 6. Flowchart for the procedure for finding shifting angle with minimum cogging torque.

Table 4. Tabulation of results of cogging torque with variation of magnet shifting from 1° to 8°.

Shifting Angle (°)	A-B		B-C		C-D		D-A		Cogging Torque (Nm)
	Angle (°)	Space (mm)	Angle (°)	Space (mm)	Angle (°)	Space (mm)	Angle (°)	Space (mm)	
1°	91	5.56	89	4.44	91	5.56	89	4.44	0.64
	89	4.44	91	5.56	89	4.44	91	5.56	0.42
	89	4.44	89	4.44	91	5.56	91	5.56	0.41
	91	5.56	91	5.56	89	4.44	89	4.44	0.61
2°	92	5.58	88	4.42	92	5.58	88	4.42	0.89
	88	4.42	92	5.58	88	4.42	92	5.58	0.34
	88	4.42	88	4.42	92	5.58	92	5.58	0.82
	92	5.58	92	5.58	88	4.42	88	4.42	0.95
3°	93	6.68	87	3.32	93	6.68	87	3.32	0.48
	87	3.32	93	6.68	87	3.32	93	6.68	0.8
	87	3.32	87	3.32	93	6.68	93	6.68	0.16
	93	6.68	93	6.68	87	3.32	87	3.32	0.25
4°	94	7.24	86	2.76	94	7.24	86	2.76	1.2
	86	2.76	94	7.24	86	2.76	94	7.24	0.22
	86	2.76	86	2.76	94	7.24	94	7.24	0.56
	94	7.24	94	7.24	86	2.76	86	2.76	1.3
5°	95	7.8	85	2.2	95	7.8	85	2.2	2.1
	85	2.2	95	7.8	85	2.2	95	7.8	0.56
	85	2.2	85	2.2	95	7.8	95	7.8	0.5
	95	7.8	95	7.8	85	2.2	85	2.2	2.5
6°	96	8.38	84	1.62	96	8.38	84	1.62	1.7
	84	1.62	96	8.38	84	1.62	96	8.38	0.33
	84	1.62	84	1.62	96	8.38	96	8.38	0.39
	96	8.38	96	8.38	84	1.62	84	1.62	3.4
7°	97	8.95	83	1.05	97	8.95	83	1.05	2.9
	83	1.05	97	8.95	83	1.05	97	8.95	0.68
	83	1.05	83	1.05	97	8.95	97	8.95	1.5
	97	8.95	97	8.95	83	1.05	83	1.05	2.4
8°	98	9.52	82	0.48	98	9.52	82	0.48	2.1
	82	0.48	98	9.52	82	0.48	98	9.52	0.7
	82	0.48	82	0.48	98	9.52	98	9.52	0.56
	98	9.52	98	9.52	82	0.48	82	0.48	4

Table 5 shows the comparison of the 3D FEA results of cogging torque when shifting the magnetic pole angle from 1° to 8°. The graphical representation of variation in the cogging torque with magnet shifting is shown in Figure 8. From the above comparison, when the shifting angle is 3°, the lowest cogging torque of 0.16 Nm is obtained. The base rotor has a cogging torque of 0.64 Nm. Compared with the base rotor, the new asymmetrical rotor structure exhibits a 75% reduction in cogging torque.

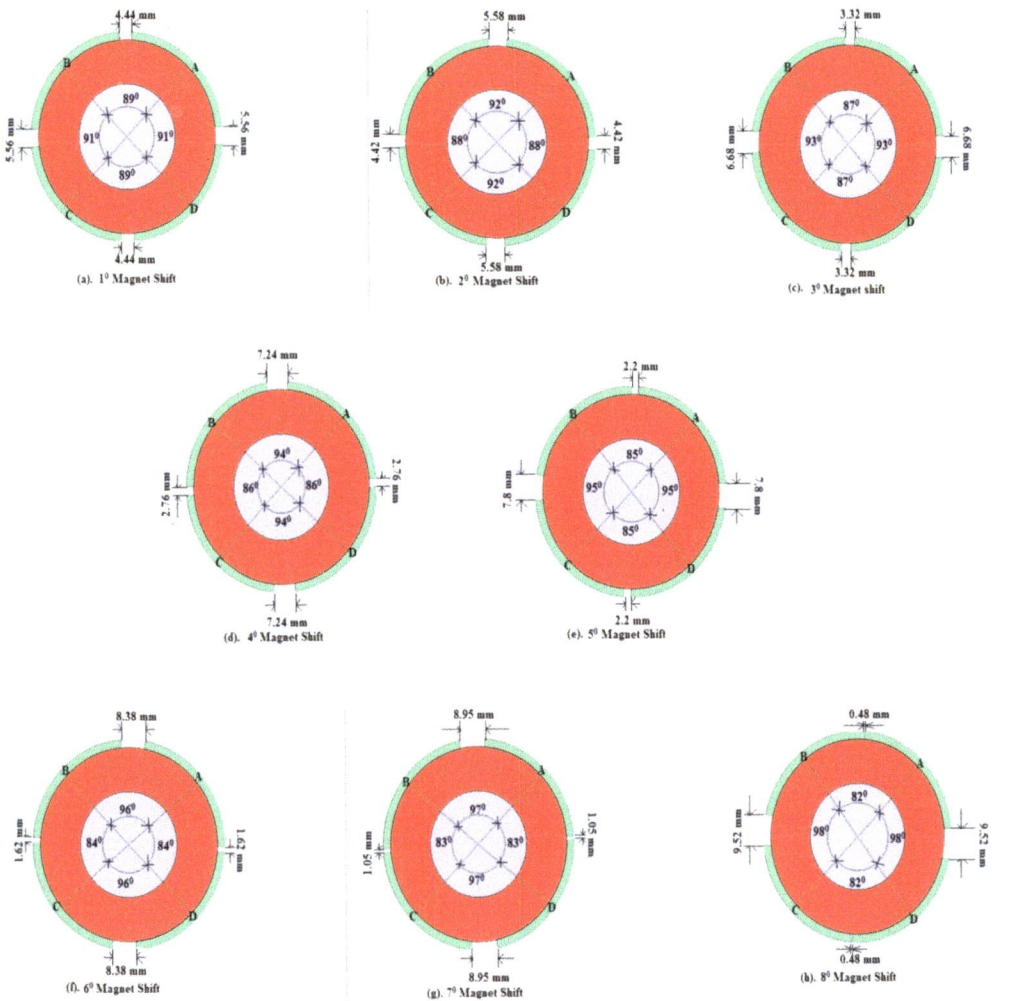

Figure 7. Rotor magnet shift from 1° to 8°.

Table 5. Comparison of 3D FEA results for cogging torque.

Sl. No.	Shifting Angle (°)	Angle between Magnets (°)				Cogging Torque (Nm)
		A-B	B-C	C-D	D-A	
1	1	89	91	89	91	0.41
2	2	92	88	92	88	0.34
3	3	87	93	87	93	0.16
4	4	94	86	94	86	0.22
5	5	85	95	85	95	0.5
6	6	96	84	96	84	0.33
7	7	97	83	97	83	0.68
8	8	82	98	82	98	0.56

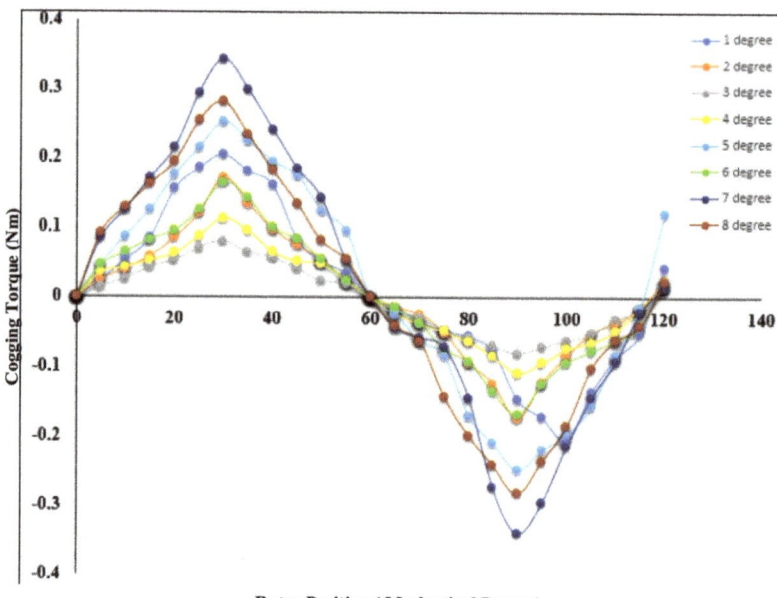

Figure 8. FEA results of cogging torque with varying shifting angle.

One of the main reasons for the existence of cogging torque is the change in magnetic flux density. Assessment of flux density and assessment of flux lines are two pivotal steps in FEA. Whenever the magnet displacement is applied to the rotor, the flux density is lower than that of the symmetrical structure. Though the flux density remains the same in many regions of the BLDC motor, the maximum flux density, indicated by yellow color, is attained in some parts of the stator (in addition to the permanent magnets). Red areas indicate undesirably high flux density, which may result in hot spots that could damage the motor. When applying magnet displacement in the rotor, the flux density decreases. This causes a reluctance to change that is comparatively better than the existing method, resulting in a reduction in cogging torque. Figures 9 and 10 show the flux plot distribution in symmetrical and asymmetrical rotors.

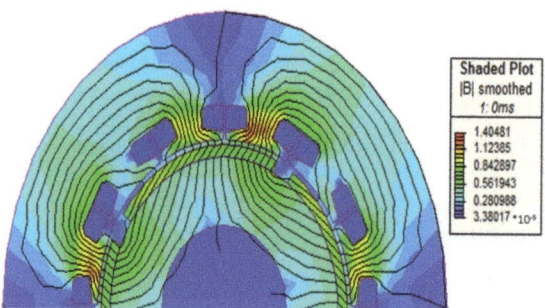

Figure 9. Flux plots in symmetrical rotor.

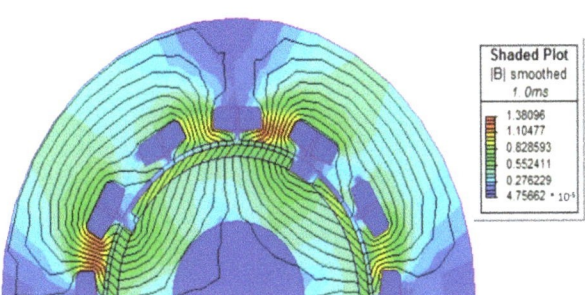

Figure 10. Flux plots in the asymmetrical rotor.

Figure 11a shows the symmetrical rotor structure with an angle difference of 90°, Figure 11b shows the asymmetrical rotor structure with 3° magnet shifting. There is a small difference in the spacing of the permanent magnet. In the symmetrical design, all the rotor magnets are placed equally at a distance of 5 mm from one another, and in the asymmetrical structure, the spacing between magnets A and B is 3.32 mm, and the spacing between magnets B and C is 6.68 mm. Table 6 shows a comparison of the cogging torque between the symmetrical and asymmetrical rotors.

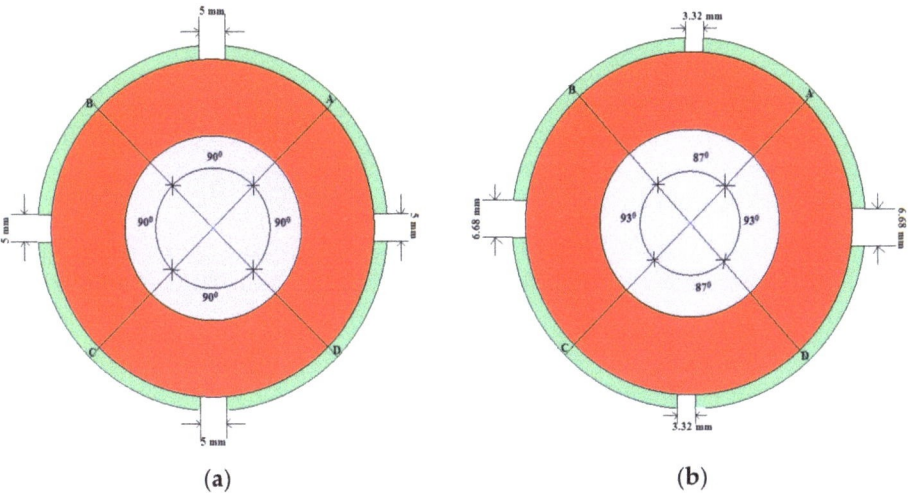

Figure 11. Compared Rotor structure (**a**) Symmetrical rotor structure with an angle difference of 90°; (**b**) asymmetrical rotor structure with 3° magnet shifting.

Table 6. Comparison of cogging torque for the symmetrical and the asymmetrical rotor.

Rotor Type	Angle between Magnets (°)				Cogging Torque (Nm)
	A-B	B-C	C-D	D-A	
Symmetrical	90°	90°	90°	90°	0.64
Asymmetrical	87°	93°	87°	93°	0.16

Table 7 presents a comparison of cogging torque results between simulation and analytical method. When the shifting angle is 3°, both methods have almost the same results. Figure 12 shows a graphical representation of the FEA results and the analytical results.

Table 7. Comparison of simulation and analytical result.

Shifting Angle (°)	Cogging Torque (Nm)	
	Simulation Result	Analytical Result
1	0.41	0.39
2	0.34	0.3
3	0.16	0.15
4	0.22	0.18
5	0.5	0.45
6	0.33	0.2
7	0.68	0.6
8	0.56	0.51

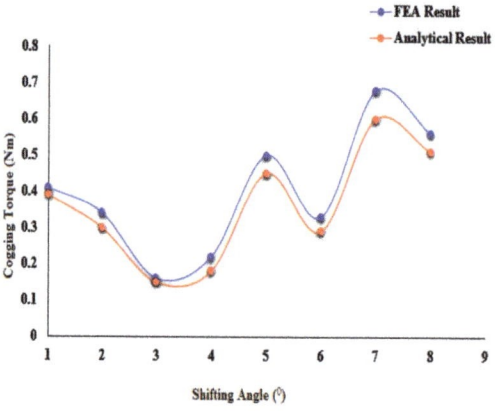

Figure 12. Comparison of FEA and analytical results.

The magnet shifting can reduce the cogging torque effectively without deteriorating the trapezoidal shape of the back-EMF. Figure 13 shows the back-EMF of the SPMBLDC motor with different magnet shifting angles. In this figure, 0° represents all four magnets being placed at an exactly 90° phase difference from one another, and is represented using blue color. The yellow color, which is very close and similar to the 0° case, phase shifts the back-EMF curve with a 3° phase shift

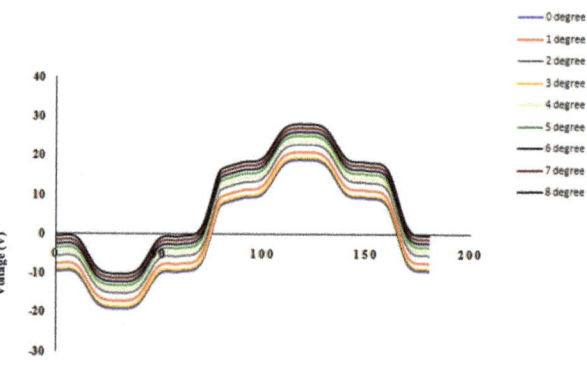

Figure 13. Comparison of back-EMF.

Figures 14–17 show the transient 3D results and the performance characteristics of symmetrical and asymmetrical rotor magnets. Even when the rotor magnetic angles are shifted, the acceleration of the BLDC motor is almost constant, and the initial speed is also maintained at a constant level, meaning that the speed of the BLDC motor increases linearly with respect to time. Figure 13 shows the time vs. speed characteristic of symmetrical and asymmetrical rotor magnets. Orange color represents the speed of the motor with the base model and blue color shows the speed of the motor with a 3-degree phase shift. From the figure, it is evident that the motor with a 3-degree phase shift is able to achieve greater speed than the base model within the specified time. Figures 14–16 present comparisons of the magnetic torque, load torque and net torque of the symmetrical magnet and the asymmetrical magnet. From the above figures, it is clear that the rotor with the 3-degree magnet shift has excellent torque vs. time characteristics when compared with the base model.

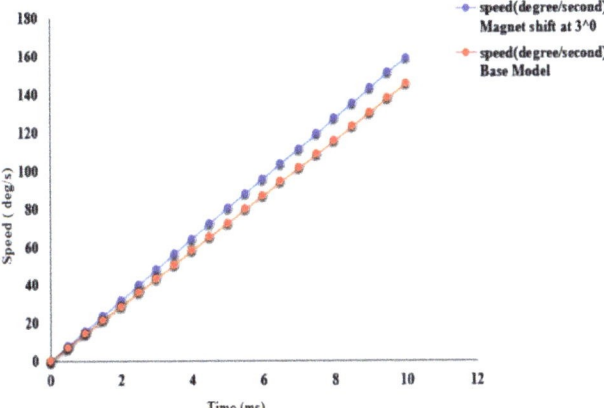

Figure 14. Speed vs. time.

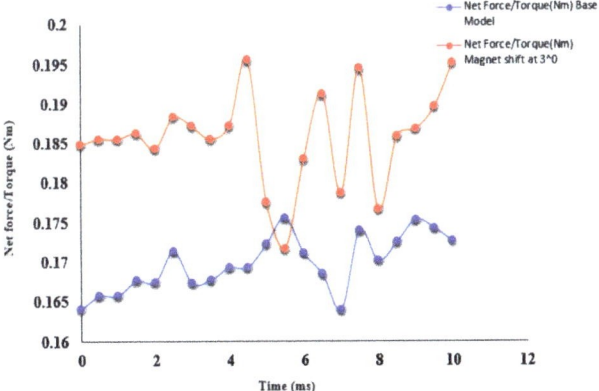

Figure 15. Magnetic force/torque vs. time.

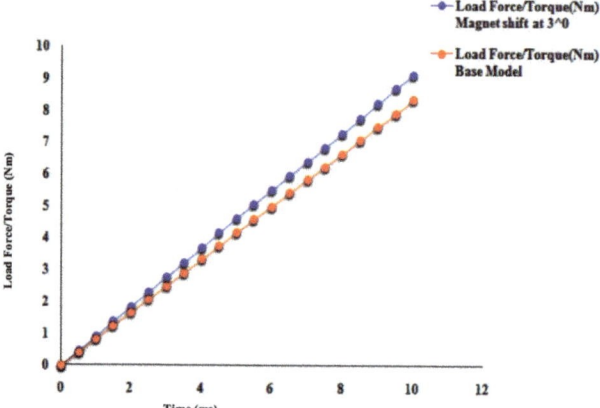

Figure 16. Load force/torque vs. time.

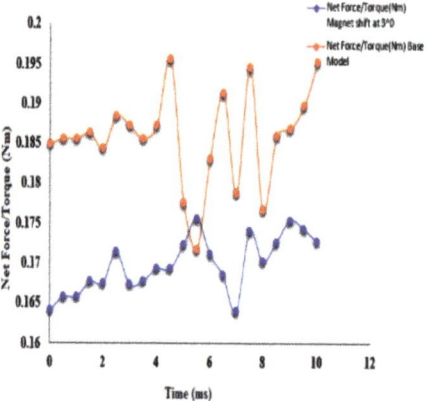

Figure 17. Load force/torque vs. time.

Figures 18 and 19 represent the cogging torque and the voltage under transient conditions. For cogging torque, the transient starts at 0.15 and increases to 0.45 Nm for a particular period before settling down to 0.16 Nm. Considering the transient scenario for voltage, it starts from 0 V and increases to 0.24 V, before after a particular period decreasing to 0.05 V and settling at 0.084 V.

In reference [29], in order to reduce cogging torque, a magnet shifting technique was adopted. The authors considered a 4 pole 12 slot machine. Table 8 shows the comparative results of the existing and proposed design.

Table 8. Comparative results of the existing and proposed design.

Parameters	Existing	Proposed
No. of slots	12	12
No. of Poles	4	4
Magnet shift angle	7.5°	3°
Pole arc	60°	63°
Cogging Torque	0.4 Nm	0.16 Nm

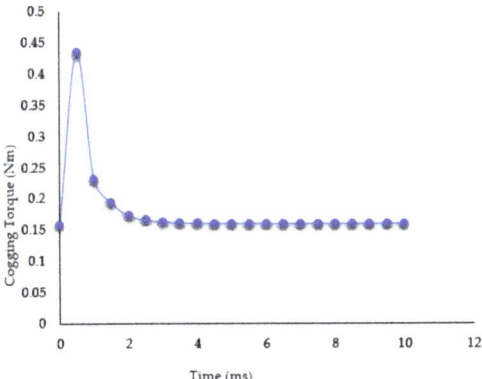

Figure 18. Transient response of cogging torque.

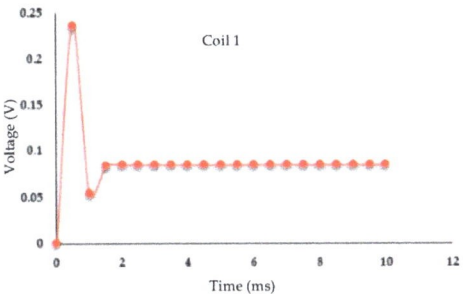

Figure 19. Transient response of voltage.

On the basis of existing work, it is clear that the new proposed model offers a 60% reduction in cogging torque.

An undesirable effect that occurs in permanent magnet motors during shaft variation is torque ripple. This is a periodic increase and decrease in output torque. In BLDC motors, cogging torque is a crucial factor contributing to torque ripple. Figure 20 presents the resulting torque ripple waveform following a 3° magnetic shift. This is the difference between the maximum torque and the minimum torque compared to the average torque [30]. The rated torque of the motor was 1.424 Nm. The maximum, minimum and average values of electromagnetic torque were 1.18 Nm, 0.6 Nm and 1.1 Nm, respectively. Therefore, the degree of torque ripple was 52.7%.

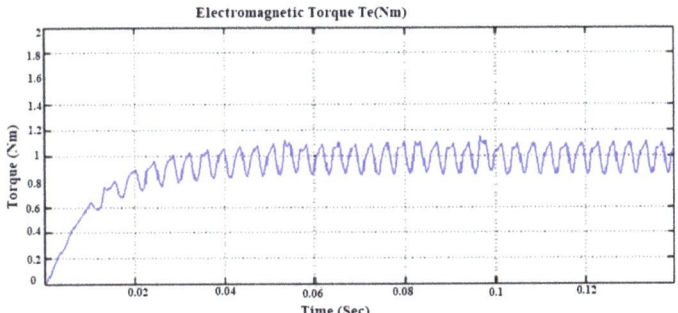

Figure 20. Torque ripple waveform with 3° magnetic shift.

6. Conclusions

A new method was proposed for the reduction of cogging torque in permanent magnet BLDC motors. In this method, a rotor with an asymmetrical magnet structure is recommended for the minimization of cogging torque. The asymmetry is achieved through the displacement of the permanent magnet. The shifting angle considered for the analysis was selected by maintaining the L/τ ratio within the permissible limit. Combinations from $1°$ to $8°$ were tried out. The magnet pole shifting caused variations in the flux line distribution and flux densities. The variation in flux density helped to achieve a reduction in cogging torque. When the shifting angle was $3°$, the minimum cogging torque was obtained. The prototype adopted for this method was analyzed using FEA, showing an immense decrease in cogging torque. A novel analytical approach was also developed in order to be able to speculate regarding the effects on cogging torque in a permanent magnet BLDC motor. The predicted cogging torque curve showed good agreement with the FEA results. The proposed method is practical and capable of achieving low cogging torque. The proposed rotor structure with an asymmetrical magnet is advantageous for the minimization of cogging torque.

Author Contributions: T.A.A. and M.A.N.D. have same contribution in design in motor and article work. All authors have read and agreed to the published version of the manuscript.

Funding: This research received no external funding.

Conflicts of Interest: The authors declare no conflict of interest.

Nomenclature

HP	Horse Power
Ph	Phase
η	Efficiency
P	No. of Poles
V	Supply Voltage
f	Frequency
Pf	Power factor
L	Stator Length
τ	Pole pitch
B_{av}	Specific magnetic Loading
ac	Specific electric Loading
Kw	Winding Factor
N	Speed
Q	kVA input
C_o	Output Coefficient
D	Stator Diameter
E_v	total energy variation
$E_{v.I}$	energy variation in iron
$E_{v.airgap}$	energy variation in airgap
$E_{v.PM}$	energy variation in PM
μ_0	permeability of air
B	magnetic flux density
α	angle of rotation of the rotor
l_m	length of the permanent magnet
l_g	length of airgap distribution
$B_{rs(\theta)}$	residual flux density along the periphery of the airgap
θ_s	Shifting angle
L_s	length of the stack
s	Slot number
R_r	Outer rotor diameter

References

1. Leitner, S.; Gruebler, H.; Muetze, A. Cogging Torque Minimization and Performance of the Sub-Fractional HP BLDC Claw-Pole Motor. *IEEE Trans. Ind. Appl.* **2019**, *55*, 4653–4664. [CrossRef]
2. Hannon, B.; Sergeant, P.; Dupre, L. Evaluation of the Torque in High-Speed PMSMs With a Shielding Cylinder and BLDC Control. *IEEE Trans. Magn.* **2018**, *54*, 1–8. [CrossRef]
3. Lee, M.; Kong, K.; Lee, M. Fourier-series-based Phase Delay Compensation of Brushless DC Motor Systems. *IEEE Trans. Power Electron* **2017**, *33*, 1. [CrossRef]
4. Kumar, B.A.; Kamal, C.; Amudhavalli, D.; Thyagarajan, T. Reformed Stator Design of BLDC Motor for Cogging Torque Minimization Using Finite Element Analysis. In *Proceedings of the 4th International Conference on Electrical Energy Systems (ICEES), Chennai, India, 7–9 February 2018*; Institute of Electrical and Electronics Engineers (IEEE): Chennai, India, 2018; pp. 481–484.
5. Nam, D.-W.; Lee, K.-B.; Pyo, H.-J.; Jeong, M.-J.; Yang, S.-H.; Kim, W.-H.; Jang, H.-K. A Study on Core Skew Considering Manufacturability of Double-Layer Spoke-Type PMSM. *Energies* **2021**, *14*, 610. [CrossRef]
6. Goryca, Z.; Różowicz, S.; Różowicz, A.; Pakosz, A.; Leśko, M.; Wachta, H. Impact of Selected Methods of Cogging Torque Reduction in Multipolar Permanent-Magnet Machines. *Energies* **2020**, *13*, 6108. [CrossRef]
7. Dini, P.; Saponara, S. Cogging Torque Reduction in Brushless Motors by a Nonlinear Control Technique. *Energies* **2019**, *12*, 2224. [CrossRef]
8. Dini, P.; Saponara, S. Design of an Observer-Based Architecture and Non-Linear Control Algorithm for Cogging Torque Reduction in Synchronous Motors. *Energies* **2020**, *13*, 2077. [CrossRef]
9. Sumega, M.; Rafajdus, P.; Stulrajter, M. Current Harmonics Controller for Reduction of Acoustic Noise, Vibrations and Torque Ripple Caused by Cogging Torque in PM Motors under FOC Operation. *Energies* **2020**, *13*, 2534. [CrossRef]
10. Doss, M.A.N.; Brindha, R.; Mohanraj, K.; Dash, S.S.; Kavya, K.M. A Novel Method for Cog-ging torque Reduction in Permanent Magnet Brushless DC Motor Using T-shaped Bifurcation in Stator Teeth. *Prog. Electromagn. Res. M* **2018**, *66*, 99–107. [CrossRef]
11. García-Gracia, M.; Romero, Á.J.; Ciudad, J.H.; Arroyo, S.M. Cogging Torque Reduction Based on a New Pre-Slot Technique for a Small Wind Generator. *Energies* **2018**, *11*, 3219. [CrossRef]
12. Hwang, M.-H.; Lee, H.-S.; Cha, H.-R. Analysis of Torque Ripple and Cogging Torque Reduction in Electric Vehicle Traction Platform Applying Rotor Notched Design. *Energies* **2018**, *11*, 3053. [CrossRef]
13. Kwon, J.-W.; Lee, J.-H.; Zhao, W.; Kwon, B.-I. Flux-Switching Permanent Magnet Machine with Phase-Group Concentrated-Coil Windings and Cogging Torque Reduction Technique. *Energies* **2018**, *11*, 2758. [CrossRef]
14. Liu, C.; Lu, J.; Wang, Y.; Lei, G.; Zhu, J.; Guo, Y. Techniques for Reduction of the Cogging Torque in Claw Pole Machines with SMC Cores. *Energies* **2017**, *10*, 1541. [CrossRef]
15. Doss, M.A.N.; Vijayakumar, S.; Mohideen, A.J.; Kannan, K.S.; Balaji Sairam, N.D.; Karthik, K. Reduction in Cogging torque and Flux per Pole in BLDC Motor by Adapting U-Clamped Magnetic Poles. *IJPEDS* **2017**, *8*, 297–304. [CrossRef]
16. Fazil, M.; Rajagopal, K.R. A Novel Air-Gap Profile of Single-Phase Permanent-Magnet Brushless DC Motor for Starting Torque Improvement and Cogging Torque Reduction. *IEEE Trans. Magn.* **2010**, *46*, 3928–3932. [CrossRef]
17. Hwang, S.-M.; Eom, J.-B.; Hwang, G.-B.; Jeong, W.-B.; Jung, Y.-H. Cogging torque and acoustic noise reduction in permanent magnet motors by teeth pairing. *IEEE Trans. Magn.* **2000**, *36*, 3144–3146. [CrossRef]
18. Park, Y.-U.; Cho, J.-H.; Kim, D.-K.; Young-Un, P. Cogging Torque Reduction of Single-Phase Brushless DC Motor with a Tapered Air-Gap Using Optimizing Notch Size and Position. *IEEE Trans. Ind. Appl.* **2015**, *51*, 4455–4463. [CrossRef]
19. Rahman, M.M.; Kim, K.-T.; Hur, J. Design and Optimization of Neodymium-Free SPOKE-Type Motor with Segmented Wing-Shaped PM. *IEEE Trans. Magn.* **2014**, *50*, 865–868. [CrossRef]
20. Han, K.-J.; Cho, H.-S.; Cho, D.-H.; Jung, H.-K. Optimal Core Shape Design for Cogging Torque Reduction of Brushless DC Motor Using Genetic Algorithm. *IEEE Trans. Magn.* **2000**, *36*, 4.
21. Hwang, K.-Y.; Rhee, S.-B.; Yang, B.-Y.; Kwon, B.-I. Rotor Pole Design in Spoke-Type Brushless DC Motor by Response Surface Method. *IEEE Trans. Magn.* **2007**, *43*, 1833–1836. [CrossRef]
22. Kim, H.-S.; You, Y.-M.; Kwon, B.-I. Rotor Shape Optimization of Interior Permanent Magnet BLDC Motor According to Magnetization Direction. *IEEE Trans. Magn.* **2013**, *49*, 2193–2196. [CrossRef]
23. Lee, S.-K.; Kang, G.-H.; Hur, J.; Kim, B.-W. Stator and Rotor Shape Designs of Interior Permanent Magnet Type Brushless DC Motor for Reducing Torque Fluctuation. *IEEE Trans. Magn.* **2012**, *48*, 4662–4665. [CrossRef]
24. Doss, M.A.N.; Jeevananthan, S.; Dash, S.S.; Jahir Hussain, M. Critical Evaluation of cogging torque in BLDC Motor with various Techniques. *Int. J. Autom. Control.* **2013**, *7*, 135–146. [CrossRef]
25. Doss, M.A.N.; Sridhar, R.; Karthikeyan, M. Cogging torque Reduction in Brushless DC Motor by Reshaping of Rotor Magnetic Poles with Grooving Techniques. *Int. J. Appl. Eng. Res.* **2015**, *10*, 36.
26. Zhenhong, G.; Liuchen, C.; Yaosuo, X. *Coggingtorqueof permanent Magnet Electric machines: An Overview*; IEEE: St. John's, NL, Canada, 2009; ISSN 0840-7789.
27. Chang, L.; Eastham, A.R.; Dawson, G.E. Permanent magnet Synchronous Motors: Finite Element Torque calculations. In Proceedings of the Conference Record of the IEEE Industry Applications Society Annual Meeting, San Diego, CA, USA, 1–5 October 1989; pp. 69–73.
28. Yang, Y.; Wang, X.; Zhang, R.; Ding, T.; Tang, R. The Optimization of Pole Arc Coefficient to Reduce Cogging torque in Surface-Mounted Permanent Magnet Motors. *IEEE Trans. Magn.* **2006**, *42*, 4.

29. Breton, C.; Bartolome, J.; Benito, J.A.; Tassinario, G.; Flotats, I.; Lu, C.W.; Chalmers, B.J. Influence of Machine Symmetry on Reduction of Cogging Torque in Permanent-Magne Brushless Motors. *IEEE Trans. Magn.* **2006**, *36*, 5.
30. Doss, M.A.N.; Mohanraj, K.; Kalyanasundaram, V.; Karthik, K. Reduction of Cogging Torque by Adapting Bifurcated Stator Slots and Minimization Of Harmonics And Torque Ripple in Brushless DC Motor. *Int. J. Power Electron. Drive Syst. (IJPEDS)* **2016**, *7*, 781. [CrossRef]

Article

Optimal Dynamic Scheduling of Electric Vehicles in a Parking Lot Using Particle Swarm Optimization and Shuffled Frog Leaping Algorithm

George S. Fernandez [1], Vijayakumar Krishnasamy [1], Selvakumar Kuppusamy [1], Jagabar S. Ali [1,2], Ziad M. Ali [3,4], Adel El-Shahat [5,*] and Shady H. E. Abdel Aleem [6]

1. Department of Electrical and Electronics Engineering, SRM University, Kattankulathur 603-203, India; George.electrix@gmail.com (G.S.F.); kvijay_srm@rediffmail.com (V.K.); Selvakse@gmail.com (S.K.); srmmjs@gmail.com (J.S.A.)
2. Renewable Energy Lab (REL), Department of Communication and Networks, College of Engineering, Prince Sultan University (PSU), Riyadh 11586, Saudi Arabia
3. Electrical Engineering Department, College of Engineering at Wadi Addawaser, Prince Sattam Bin Abdulaziz University, Al-Kharj 11991, Saudi Arabia; dr.ziad.elhalwany@aswu.edu.eg
4. Electrical Engineering Department, Aswan Faculty of Engineering, Aswan University, Aswan 81542, Egypt
5. Department of Electrical and Computer Engineering, Georgia Southern University (GSU), Statesboro, GA 30460-7995, USA
6. Power Quality Department, ETA Electric Company, El Omraniya, Giza 12111, Egypt; engyshady@ieee.org
* Correspondence: aahmed@georgiasouthern.edu

Received: 22 September 2020; Accepted: 1 December 2020; Published: 3 December 2020

Abstract: In this paper, the optimal dynamic scheduling of electric vehicles (EVs) in a parking lot (PL) is proposed to minimize the charging cost. In static scheduling, the PL operator can make the optimal scheduling if the demand, arrival, and departure time of EVs are known well in advance. If not, a static charging scheme is not feasible. Therefore, dynamic charging is preferred. A dynamic scheduling scheme means the EVs may come and go at any time, i.e., EVs' arrival is dynamic in nature. The EVs may come to the PL with prior appointments or not. Therefore, a PL operator requires a mechanism to charge the EVs that arrive with or without reservation, and the demand for EVs is unknown to the PL operator. In general, the PL uses the first-in-first serve (FIFS) method for charging the EVs. The well-known optimization techniques such as particle swarm optimization and shuffled frog leaping algorithms are used for the EVs' dynamic scheduling scheme to minimize the grid's charging cost. Moreover, a microgrid is also considered to reduce the charging cost further. The results obtained show the effectiveness of the proposed solution methods.

Keywords: charging cost; dynamic charging; economics; electric vehicles; optimization; parking lots; static charging

1. Introduction

Many research works have been presented in the literature to overcome the issues related to the electric vehicles' (EVs) scheduling at parking lots (PLs), such as the number of charging points, time-varying electricity price, the capacity of chargers, and charging limit. However, few works addressed advanced technologies for online booking and location finding [1]. In this regard, the research works presented in the literature are broadly focused on three major categories: (i) EV battery charging technology, (ii) charging scheduling schemes, and (iii) charging station (CS) recommendation methods. In global environmental pollution, the transport sector has a significant role due to fossil fuel usage. Nowadays, a non-fuel or a partial fuel-based vehicle is emerging due to low fuel consumption, no environmental pollution, and reduction of greenhouse emissions, etc. [2,3]. Most countries migrate from fuel to electrical-based transport systems (EV-based systems), and more research and development is initiated in this direction. In Figure 1, the typical schematic diagram charging system is shown. By changing conventional vehicles to EVs, the electric power supply has to be maintained with power quality. However, in practice, a large amount of EV charging degrades the electric power system's performance due to unexpected demand, overload of transformers, and grid stability issues [4,5].

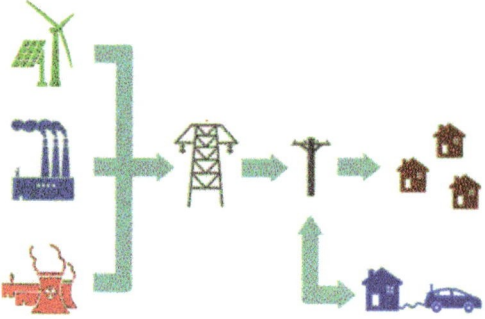

Figure 1. Typical electric vehicle (EV) charging system.

The EV charger is one of the main components in determining the recharging time of the battery. In general, the commonly used EV chargers are classified into four types:

- Based on the location in which this type of charger is available in the EV itself.
- Based on the power level, the charger is not equipped with the EV and is available separately. The onboard chargers available are fitted with the EVs, but the charger increases the EV's weight due to the additional circuit. The off-board charger requires a dedicated CS to facilitate charging [6]. In general, the off-board chargers are advantageous compared to the onboard chargers in terms of being placed in public-accessing areas like a parking lot, bus stop, etc. They can also be classified into slow, medium, and fast charging based on the power ratings. However, the fast-charging facility is used to complete the charging in a short time [7,8] compared to the other two types. Four different charging levels are classified by international standards such as the Society of Automotive Engineers (SAE), Electric Power Research Institute (EPRI), and International Electro-technical Commission (IEC) [9]. The different types of EV chargers can be found in Ref. [9].
- Charges based on the source is classified as (i) constant current, in which the input current is constant, (ii) constant voltage, in which the input current is variable through the charging, and (iii) hybrid constant current and voltage to allow fast charging without risk of over-charging.
- Wireless-charging schemes avoid the usage of cables, but they have low-efficiency. However, ongoing research works are under development to improve the efficiency of these wireless charging schemes.

The EV charging loads can double the average household electricity consumption and makes the user pay more. On the other hand, it worsens the distribution network during peak time. Hence, controlled EV charging is a prominent solution to minimize grid disturbances [10]. Most plug-in hybrid EV (PHEVs)/EV charging is predicted to occur in public CS. During the peak load period, the controlled EV charging is used to minimize the grid disturbances and charging costs [11,12]. Figure 2 shows an EV charging at a parking lot.

Figure 2. EV charging at a parking lot.

Controlled EV charging is classified as centralized charging and decentralized charging. The aggregator or a standard operator [13] will control individual PHEVs and make a universal control for cost reduction in the centralized coordination scheme. However, this scheme is not advisable for the customers who do not want any third party to control their EV usage and electric power consumption. Under the decentralized charging scheme, there is no restriction on using the charging point. The EVs can occupy the charging point directly until the battery is charged sufficiently. A PL with 2 to 6 chargers is highly recommended due to space and cost constraints [14]. Some demand response (DR) programs are considered for a smart grid to control the peak demand and minimize EV charging costs. A new intelligent load management scheme is proposed to reduce the charging cost. This coordinated charging scheme is implemented by controlling several EVs charging and taking load profiles of the residential area into account. This is investigated in multiple residential distribution systems with EVs. The charging time is shifted to midnight to minimize the charging cost and peak load without using a storage device. However, the proposed scheme does not discuss the charging infrastructure, uncertain arrival, and EVs [15]. A real-time power management program is recommended, and optimal scheduling is implemented using a genetic algorithm (GA). However, the uncertain arrival of EVs and real-time pricing are not considered [16,17]. A distributed DR program is proposed to manage EV charging demand and minimize the charging cost in a smart grid [18]. In this method, the forecasted electricity price is shared with the customer. Besides, several DR programs are presented in the energy and reserve market for optimal EVs schedule [18,19]. By deciding the charging and discharging time from each battery optimally, the PL's profit is maximized [20,21]. The suggested approach is compared with the time of use (TOU), critical demand price, and emergency DR programs. The choice of charging and discharging of EVs is completed at a particular time sequence with an unvarying rate. An aggregator supported centralized EV charging is proposed to minimize the overall purchasing cost of electricity. However, the provided solution requires a high-level communication infrastructure. Also, it is assumed that CSs have unlimited electric power to charge a considerable number of EVs together, which is not easy to implement in a real-time scenario. The charging cost variation in the CSs provides the EV users to choose between charging time. Considering this, a charging model is proposed for a PL equipped with a solar and energy storage system (ESS) [22]. This model also includes the PL's profit maximization, the capacity of distributed generation (DG) and ESS, PL's investment choices, and cost to charge the EVs. However, the basic first-in-first serve (FIFS) method is used for charging the EVs.

The EV charging scheduling is presented in [23] with ESS from a power market perspective. The aggregator considered the day ahead and actual market price and involves the energy trade. The optimal charging improves the aggregator's revenue, and it can be further enhanced with ES's support. However, it is assumed that the EV charging demand is known. A charging scheme considering a real-time scenario to minimize the EV charging cost is presented [24]. This scheme includes EV demand, which varies with power tariff and load reduction requests from service providers. The proposed system operates with a dynamic tariff provided by the operator. The EV charging schedule is determined by turning on/off each charger available in the PL. An optimal charging scheduling scheme in an office PL is proposed in [25] using a two-stage relative dynamic program. The EV arrival pattern is modeled using the Poisson process. The Poisson process is a model for a series of discrete events where the average time between events is known, but the exact timing of events is random. The primary goal of optimal scheduling is to reduce the cost of EV charging. A penalty cost is also considered if the PL does not provide the requested power. A game-theoretic approach is used to schedule the EV charging [26]. The proposed method considers the variation of hourly energy costs to minimize the charging cost. However, the vehicle charging demand of EVs is very low. Optimal resource sharing to minimize the charging cost is proposed for the municipal PL [27]. A large number of EVs are scheduled in a PL by using a distribution algorithm. The available state of charge (SOC), the time required to reach full battery capacity, and the utility cost are considered to minimize the charging cost. The charging rate is regarded as a continuous variable. A two-layered parking lot for the EV recharging scheme is proposed in [28] to minimize the EV charging cost. The proposed system is compared with the basic charging scheduling scheme, such as FIFS and early deadline first (EDF). However, the recommended procedures require high-level communication network support between the users and aggregators for making optimal scheduling to minimize the charging cost.

Also, optimal scheduling for EVs with random arrival time is proposed in [29]. The battery's capacity, available SOC, charging interval, arrival and departure time, and charging cost are considered for the charging cost minimization. The price is kept constant throughout the charging locations. The EVs are grouped and managed by a local controller. The predicted demands are sent from the central controller to a local controller—the local controller schedules the EV based on the optimization algorithm to charge or discharge the EV. A smart charging method using a third-party agent is presented in [30]. The required power to charge the EV is shared with the aggregator. The aggregator considers the power allocated by the distribution system operator (DSO) and transmission system operator (TSO) to make optimal scheduling. However, this method only finds a monotype of charging. However, implementing this scheme in real-time is not economical due to technical challenges. In [31], smart charging is proposed to enable optimal EV charging. Two algorithms are developed to minimize the charging cost. By analyzing the predicted electricity cost, dynamic programming is designed to obtain the charging cost. However, the forecasted driving profile and power requirements are not always accurate. In [32], an agreement-based approach is proposed to minimize the charging cost. The EV user needs to sign an agreement with the aggregators for charging the EV. The drawback of this method is that the users are forced to charge the EV for a particular time every day. A flexible EV scheduling scheme is developed to minimize the charging cost [33]. In this scheme, the customer details are shared with several operators such as charging service provider (CSP), DSO, and a retailer. The system operator forecasts the EV charging load to find an optimal schedule. However, predicting the EV load is not always accurate. Furthermore, the privacy of the customer is affected. In [34], a price-response based EV scheduling method is proposed using modern communication infrastructure. A base-level aggregator and a central aggregator are involved in the EV scheduling for charging cost minimization. In this scheme, it is assumed that the demand for EVs, plug-in time, and charging time of EVs are known to the base level aggregator. In [35], a charging scheduling scheme is proposed by considering vehicle uncertainty. Bidirectional communication is used for monitoring and controlling the data exchange between aggregators and users. The global aggregator decides the charging management of EVs in the CSs. In [36], a model is developed for the PL operator to charge EVs in a deregulated power market. The objectives are to increase the service provider's revenue and the revenue from renewable resources. In this model, the service provider utilizes the EVs by discharging the power left at the battery. Because of battery power discharging, the expected SOC of the EV may not be reached when the EV customer wants to depart from the PL. A charging scheduling scheme is developed to minimize the EV charging cost in a PL. A bi-level approach to bid the electricity price is introduced between the aggregator and DSO. This method considers the generation limit with the uncertainty of wind power and charging demand [37]. An optimal charging scheduled is proposed in [38] to minimize the charging cost of PL. The proposed scheme avoids grid disturbance at the distribution level.

An optimal energy management scheme is proposed in a commercial PL [39]. The energy management scheme reduces the cost of the PL operator with respect to the TOU tariff. However, it is assumed that the arrival time of EV and demand of EV is already known to the PL operator. Smart charging management for EVs in a PL is carried out to reduce the charging cost [40]. The PL is equipped with a photovoltaic system and an ESS. The proposed method minimizes the charging cost of the PL. However, the charging scheduling for the un-appointed EVs is not considered. A smart scheduling approach is proposed for the EVs to minimize the cost by reducing the waiting time with a limited charging infrastructure [41]. A simulation is carried out for the EVs to find the CS location on a highway. However, it focused only on travel time and did not consider the EV's energy consumption, varying with the EV speed. An optimal charging schedule for EVs is proposed in [42]. The charging cost variation was calculated by considering the uncertain arrival and departure of EVs. An aggregator controlled dynamic scheduling scheme is proposed in [43] to minimize the charging cost. The objective is derived from the total cost and a penalty cost to the operator if the charging is not completed before the user's timeline. An optimal centralized EV charging scheduling is developed in [44] for minimizing the overall charging cost. But it is assumed that the CS is having an unlimited number of charging points to avoid queuing. Also, if the EV is plugged in for charging, it cannot be plugged out until the battery is fully charged. However, in a dynamic tariff, the energy cost is variable, and hence the scheduling cannot be shifted when the energy price is low. In [45], a cost-effective charging scheme is proposed by considering the output power from a photovoltaic (PV) system. The user preferences, such as charging time, required demand, parking time, etc., are included in this scheme to minimize the charging cost. An optimal day ahead charging schedule is proposed in [46] to reduce the charging cost. The aggregator considers the demand for EVs and the energy price for the optimal scheduling of EVs. A transactive control method [47] proposed two-stage optimal scheduling of EVs for charging cost minimization. An aggregator collects the day-ahead electricity price and the real-time electricity price for the charging. It is assumed that the users give their exact travel patterns for the next day and reserves a charging slot. The customers with flexibility in EV charging time obtain benefits, whereas the other EV users do not benefit. An optimal charging schedule to minimize the charging cost through vehicle-to-grid (V2G) technology is proposed in [48].

The aggregator considers the charging and discharging of multiple EVs in the CSs and minimizes the overall cost. However, frequent charging and discharging will affect the battery's life. A risk-aware day ahead EV charging scheduling scheme is proposed in [49]. This scheme reduces the difference between the actual and forecasted EV load and allocates the power to optimize the cost. The change in forecasted EV load varies with the unexpected arrival of EV. However, the uncertain arrival of more EVs makes the computation more complex. In [50], a charging schedule for EVs is proposed by considering EV users' behavior and output from a PV system. The objective was to minimize the overall charging cost. The proposed scheme estimates the EVs demand as very low. An optimal scheduling scheme was proposed in [51] for EV charging in a CS to minimize charging costs. The service provider calculates the expected scheduling demand and actual scheduling demand. If the actual demand is more than the expected demand, then the service provider cannot meet some EVs' demands. In [52], optimal scheduling is used to minimize the charging cost of the CS operator. A central operator controls the CSs, in which the central operator receives the demand requests for every hour from the CSs. This may create computational complexity in the method.

An optimal scheduling scheme is proposed for EV charging in the CSs [53] to minimize the charging cost. Various renewable energy sources such as wind, solar, and local energy storage devices are used to charge the EVs. The basic FIFS scheme is used for the EV charging. Optimal cost-based scheduling is proposed in [54] by considering renewable energy. This scheme optimizes the EV charging cost by considering the energy price, renewable energy, the arrival, and departure time of EVs. An agent-based decentralized optimal charging scheme is proposed for minimizing the cost [55]. A two-way communication scheme is introduced between the customers and the operators for sharing information such as demand, energy price, etc. A dynamic stochastic optimization method is proposed in [56] to minimize the charging cost. The users have to request the aggregator in advance by using a communication network. The aggregator allocates electric power based on the energy price. A two-stage economic operation of a PL equipped with a microgrid is proposed in [57] to reduce the charging cost. The forecasted electricity price determines the PL operation for the next 24 h. A dynamic algorithm is proposed for a coordinated charging between the EV user and the aggregator in [58]. The proposed algorithm generates the next day's EV schedule based on an EV's previous days driving pattern. The charging schedule suggested by this scheme minimizes the charging cost. However, the actual driving pattern differs from the expected driving pattern. The profit maximization of CS is developed by using an admission control program [59]. The EV demand is modeled from past historical data, and the EVs are suggested to charge in any of the CS located nearby. An optimal cooperative charging strategy is developed for the smart CS to minimize the overall charging cost of the CS [60]. The available battery power and the demand is shared with the aggregator for optimal scheduling.

The literature discussed above shows various EV charging scheduling schemes that can benefit the CS owners. Many researchers solve the charging scheduling for PL cost minimization. However, many of the researchers focus on the fixed power range of chargers and vehicles. Besides, the charging limit of the PL is also not considered. Accordingly, in this paper, the EV charging schedule to minimize the charging cost of the PL is investigated considering controlled and uncontrolled EV charging. The nature of EV charging in the PL is different. In some cases, the charging time is fixed and flexible, whereas in some cases, the charging time is variable; an EV may come with and without appointments. This different nature of PL charging methods motivated the authors to work on an economical charging schedule to minimize the PL's charging cost.

The scheme proposed in this work is more suitable for the PLs with a limited number of charging points. The PL can accommodate more EVs for charging without enhancing the PL infrastructure. The advanced booking may help the customers to avoid unwanted waiting time at the PLs. Finally, the potential of renewable energy sources in the PL is considered to reduce the charging cost. The PL operator suffers from many uncertainties in terms of EVs' energy demand, uncertainties in electricity cost, different arrival and departure time of EVs, and resources available at the PL. Hence, a dynamic charging scheme is considered the main objective of this work to minimize the PL's electricity purchase cost. In general, the arrival and departure of EVs in a PL are unpredictable. The customer has their optional preference to charge the EVs either with a prior booking or without booking. By considering the uncertain arrival of EVs, the dynamic scheduling scheme is analyzed based on the FIFS method, particle swarm optimization (PSO), and shuffled frog leaping algorithm (SFLA).

The rest of the paper is organized into five sections. The configuration of the system studied is introduced in Section 2. The problem formulation for the scheduling is given in Section 3, in which the objective function, constraints, and the three solution methods used in this work are presented in detail. The results obtained are presented and discussed in Section 4, and conclusions are presented in Section 5. Possible future works are presented in Section 6.

2. System Studied

Most of the people living in apartments are used to charge the EVs at PL located in the office or shopping complex, etc. [61]. The EV user tries to charge their vehicle in the PL during the parking time. So, the EVs are recharged when the user is engaged with other work. This work assumes that the EV's demand is to charge the battery to its full capacity. With the available information, the PL operator can utilize the developed charging scheme to develop an optimal scheduling to minimize the charging cost. So, based on the charger limits, the number of EV charging requests can be accepted. The charging limit is set to 61.5 kWh (a 30 kWh, 20 kWh, and 11.5 kWh charger).

The primary constraint that limits the number of EVs charging in a PL is the charging capacity. Furthermore, 20 EVs with different capacities and demand levels are considered for the scheduling, as shown in Table 1. The parking time of EVs is given in Table 2. The grid cost considered from the European Power Exchange (EPEX) spot is given in Table 3 [62].

Table 1. EV data.

EV (ID Number)	Capacity (kW)	Available SOC (%)	EV (ID Number)	Capacity (kW)	Available State of charge (SOC) (%)
1	17.6	8	11	24.0	29
2	23.0	25	12	27.0	38
3	16.5	10	13	16.0	40
4	24.0	14	14	17.6	33
5	27.0	19	15	23.0	30
6	16.0	23	16	16.5	27
7	24.0	28	17	30.0	16
8	30.0	12	18	17.3	18
9	17.3	30	19	32.0	34
10	32.0	35	20	16.5	25

Table 2. EV parking profile.

ID/Timeslot	1	2	3	4	5	6	7	8
1	In	-	-	Out	-	-	-	-
2	-	In	-	-	Out	-	-	-
3	-	-	In	-	-	Out	-	-
4	-	-	-	In	-	-	-	Out
5	In	-	-	-	-	-	Out	-
6	-	-	In	Out	-	-	-	-
7	-	In	-	-	Out	-	-	-
8	-	In	-	-	-	-	-	Out
9	-	-	-	-	In	-	Out	-
10	-	-	-	In	-	-	Out	-
11	-	-	-	-	-	In	-	Out
12	-	-	-	-	In	-	Out	-
13	-	In	Out	-	-	-	-	-
14	-	-	-	-	In	Out	-	-
15	-	-	-	-	In	Out	-	-
16	-	-	In	-	Out	-	-	-
17	-	-	In	-	-	-	-	Out
18	-	-	-	In	-	Out	-	-
19	-	In	-	-	-	-	-	Out
20	In	-	-	-	-	-	Out	-

Table 3. Dynamic price tariff.

T (h)	1	2	3	4	5	6	7	8
Cost (€ct/kW)	7.9	7.4	7.2	6.9	6.9	7.2	10.5	24.9

In this work, a PL equipped with a microgrid (MG) is considered, as in [63]. A microturbine, five PV units, and a wind turbine are considered in the MG. The power generation limits of the renewable sources are shown in Table 4, and the cost coefficients are given in Table 5. The charging of EVs from the grid or the MG depends on the electricity price.

Table 4. Power limits of the distributed generation (DG) sources.

Type	The Lower Limit (kW)	The Upper Limit (kW)
Microturbine	6.0	30.0
Wind turbine	3.0	15.0
PV_1	0.0	3.0
PV_2	0.0	2.5
PV_3	0.0	2.5
PV_4	0.0	2.5
PV_5	0.0	2.5

Table 5. Cost coefficient of the DGs.

Type	a_i	b_i	c_i
Microturbine	0.01	5.10	46.10
Wind turbine	0.01	7.80	1.10
PV_1	0.01	7.80	1.00
PV_2	0.01	7.80	1.00
PV_3	0.01	7.80	1.00
PV_4	0.01	7.80	0.10
PV_5	0.01	7.80	1.20

The 24 h microgrid power price is calculated from the renewable sources' cost coefficients given in Table 5, and the microgrid price shown in Table 6. The MG power price is lower than the grid power price for all 8 slots. However, the EV discharging scheme such as vehicle-to-grid (V2G) is not considered in this work. In many research works, common types of EV with the same battery capacity are considered. Also, the demand for EVs is minimum. In this work, EVs' different capacities with different EV demand and multiple charging slots are considered. The problem formulation for the scheduling is given in Section 3.

Table 6. Microgrid price.

T (h)	1	2	3	4	5	6	7	8
Cost (€ct/kW)	6.1	5.7	5.7	5.5	5.9	5.9	7.9	7.3

3. Problem Formulation

The optimal scheduling problem is formulated to minimize the charging cost by considering the variation in electricity price and the allocation of chargers to various EVs and the charging limit of the PL. In this dynamic scheme, the scheduling is undertaken for every time slot because EVs may come randomly. Each timeslot is considered 60 min due to the hourly change in the electricity price.

3.1. Objective Function

The objective function is to reduce the electricity purchase cost from the grid, keeping in mind that it will be more useful for the economic operation of the PL if the operator assigns the EVs to the chargers in an optimal manner considering the electricity price and charging limit. The objective function is to minimize the charging cost as given in Equation (1).

$$C(t) = \sum_{t=1}^{T} \left(\sum_{i=1}^{NF} Ci(t)R_i + \sum_{j=1}^{NM} Cj(t)R_j + \sum_{k=1}^{NS} Ck(t)R_k \right) \tag{1}$$

where $C(t)$ is the total purchase cost of electricity to charge all the EVs. NF is the number of fast chargers, NM is the number of medium chargers, and NS is the number of slow chargers. T is the total time to charge all EVs, which is calculated using Equation (2).

$$T = \sum_{n=1}^{N}\left(\sum_{i=1}^{NF}\left(\frac{V_c^n - SOC(n)}{P_{ifc}}\right) + \sum_{j=1}^{NM}\left(\frac{V_c^n - SOC(n)}{P_{jmc}}\right) + \sum_{k=1}^{NS}\left(\frac{V_c^n - SOC(n)}{P_{ksc}}\right)\right) \qquad (2)$$

where N is the total number of vehicles, V_c^n is the rated power capacity of the EV, $SOC(n)$ is the power left in the nth vehicle, P_{ifc}, P_{jmc} and P_{ksc} are the rated charging power capacity of the fast, medium, and slow chargers.

The required power R_P to charge the EV is calculated as follows:

$$R_p = V_c^n - SOC(n) \qquad (3)$$

The charging time (R) to reach 100% SOC level is given in Equation (4).

$$R = \frac{R_p}{P_c} \qquad (4)$$

where P_c is the charger rated output power.

The charging cost of each EV is calculated using Equation (5).

$$C(n) = R_p \times E_c(t) \qquad (5)$$

where $E_c(t)$ is the electricity price at time t.

3.2. Constraints

The various constraints considered in the problem are given below.
The battery of any EV that departs the PL should be charged to 100%, which is given in Equation (6).

$$SOC(n)^{dep} = SOC(n)^{max} \qquad (6)$$

The proportion of the allocated power at any timeslot to an EV should be between 0.1 and 1 as given in Equation (7). Furthermore, to ensure that all the EVs are charged to 100% of the battery capacity, the sum of all proportion should be equal to 1. The allocated power should be within the limit of all the chargers' rated output as represented in Equation (8).

$$0 \leq D^t_{power(n)} \leq 1 \qquad (7)$$

$$D^t_{power} \leq C^{lim}_{power} \qquad (8)$$

The dynamic charging scheduling is first examined by the conventional FIFS method, and then the optimization techniques, PSO and SFLA, are used for minimizing the electricity purchase cost of the PL.

The non-booked EV can be allowed based on the two following conditions:

- If the charging can be completed before the arrival of EVs with reservation.
- If the CS limit is not violated.

However, it should be noted that if these conditions are not satisfied, it will be considered an unwanted charging request for the PL operator. In such a case, the user has to decide whether to reduce the demand or extend the departure time.

3.3. Solution Methodology

The FIFS is generally used in PLs as it allows the EVs to start charging whenever they arrived. If the number EVs arrived at the CS is more than the available charging points, it creates complications in allotting the EVs to the suitable charger for proper scheduling. Time-varying electricity price impacts charging costs, and hence the operator has to consider the electricity price at each hour. So, optimal charging scheduling requires an optimization method to minimize the charging cost. Thus, PSO and SFLA techniques are used to obtain the optimal scheduling.

3.3.1. Electric Vehicle (EV) Charging Based on First-In-First Serve (FIFS) Algorithm

The FIFS scheme permits the EVs to charge if a charging point is available to use. If all the PL's charging points are occupied, then the other EVs have to wait until a charging point is available to use. In the FIFS scheduling, the EVs have to charge with the available charging point even though a better solution is available.

3.3.2. EV Charging Based on Particle Swarm Optimization (PSO)

The PSO algorithm is inspired by birds' swarm behavior flocking and fish schooling for guiding the particles to find the optimal global solution [64]. Generally, in PSO, the population particles are spread randomly and assumed to be flying in the search space. The information exchange between the particles influences the position and velocity of each particle iteratively. Based on the personal experience, each particle possesses the best solution achieved so far. A global best solution is found from the social experience of the swarm. The impact of personal best and global best is balanced by using a randomized correction factor.

In general, X_i represents the existing position of the ith particle, V_i is the velocity of the ith particle with a distance in a unit time, $Pbest$ denotes the individual best position of the ith particle (local best), and $Gbest$ represents the global best value obtained. Mathematically, the velocity and position of each particle are updated respectively, as follows:

$$V_{i,j}^{k+1} = \left(\omega \times V_{i,j}^{k}\right) + c_1\left(rand1 \times \left(Pbest_{i,j} - X_{i,j}^{k}\right)\right) + c_2\left(rand2 \times \left(Gbest_{i,j} - X_{i,j}^{k}\right)\right) \quad (9)$$

$$X_{i,j}^{k+1} = X_{i,j}^{k} + V_{i,j}^{k+1} \quad (10)$$

where $V_i(k)$ and $X_i(k)$ are the velocity and position of the ith particle at iteration k. So that:

$$V_{i,j} = \left(V_{i,1'}, V_{i,2'}, \ldots . V_{i,j'}\right) \quad (11)$$

$$X_{i,j} = \left(X_{i,1'}, X_{i,2'}, \ldots . X_{i,j'}\right) \quad (12)$$

Also, c_1 and c_2 are the coefficients of cognitive and social acceleration, which exchange the impact of the top solutions on the particle's velocity. Further, $rand_1$ and $rand_2$ are random numbers range between 0 and 1. inertia weight ω is linearly reduced from ω_{max} to ω_{min} with the iteration as given in Equation (13). Finally, the cost function, or the global best, is calculated further as shown in Equation (14).

$$\omega = \omega_{max} - \left(\frac{\omega_{max} - \omega_{min}}{k_{max}}\right)k \quad (13)$$

$$Gbest_{i,j}(k+1) = \begin{cases} Gbest_{i,j}(k) \text{ if } f\left(Pbest_{i,j}(k+1)\right) \geq f(Gbest_{i,j}(k)) \\ Pbest_{i,j}(k) \text{ if } f\left(Pbest_{i,j}(k+1)\right) < f(Gbest_{i,j}(k)) \end{cases} \quad (14)$$

The step by step procedure of implementing the PSO algorithm is given as follows:

Step 1: Initialize the number of EVs, number of chargers, capacity of EVs, and capacity of charger, and the population size. Each EV is assigned to different chargers randomly in the population, and the charging scheduling is generated.

Step 2: Assign the number of iterations as 100.
Step 3: For every random scheduling generated, calculate the charging cost, and find the *Pbest*. Assign *Pbest* as *Gbest* for the first iteration.
Step 4: For the second iteration, Equations (11) and (12) are used to provide the updated random scheduling by assigning EVs to the different chargers. Then, calculate the charging cost of each schedule. Find *Pbest* in the second iteration.
Step 5: Compare *Pbest* of the second iteration with *Gbest* of the previous iteration. If the second iteration's charging cost is lower than the previous iteration, go to Step 7.
Step 6: If the second iteration charging cost is not lower than its value in the last iteration, go to Step 4.
Step 7: Update *Gbest* from Step 5.
Step 8: Repeat the procedure until the number of iterations is completed.

3.3.3. EV Charging Based on Shuffled Frog Leaping Algorithm (SFLA)

The SFLA is a meta-heuristic or, precisely, a memetic approach motivated from frog jumping. This algorithm considers a frog group's observed behavior while finding a location with a maximum amount of food. A population of frogs is randomly assigned in the search space. The memeplexes are generated by dividing the population into several groups. The memeplexes are evolved separately in different directions within the search space. In every memeplex, the frogs are influenced by each other. This influence makes the frogs experience a memetic evolution. Hence, the memetic evolution helps the memeplexes enhance every frog's performance to achieve the goal. During the evolution, an individual frog can change the direction based on the best frog's information in a memeplex or from the population's best frog. After an individual frog has improved its position, the frog's information can be enhanced further. The memeplexes are shuffled with each other after a particular number of memetic evolution, and then the new memeplexes are generated. This improves the ability of the frogs to find the best solution within the search space.

The position of the worst frog is updated, following the expressions given in Equations (15)–(17).

$$S_i = r \times (X_b - X_w^{new}) \quad (15)$$

$$X_w^{new} = X_w^{current} + S_i \quad (16)$$

so that;

$$Si_{min} < Si < Si_{max} \quad (17)$$

where the variation of the frog's location in a single jump is Si. r is a random uniformly distributed number ranging between 0 and 1. The most and least permissible variation of the frog's location is Si_{min} and Si_{max}. The number of memeplexes is 10, the number of frogs in a memeplex is 10, the number of frogs in a sub memeplex is 10, and the population size is 100. Si_{min} and Si_{max} vary from 0.9 to 0.4, the tolerance is 0.1, and the random value ranges from 0 to 1. The step by step procedure of implementing the SFLA is given as follows:

Step 1: Initialize the number of EVs, number of chargers, capacity of EVs, demand, the capacity of chargers, etc.
Step 2: Generate the population P by randomly assigning the EVs to the different chargers. Divide the population into M number of memeplexes.
Step 3: Calculate the charging cost of each schedule, and the costs are arranged in descending order, and then the memeplexes are generated.
Step 4: Within each memeplex, calculate each scheduling's charging cost to find out the minimum charging cost and the maximum charging cost. Assign the minimum charging cost (X_b) and the highest cost as $\left(X_w^{current}\right)$. For the first iteration, assign (X_b) as the global best solution.

Step 5: For the next iteration, update the scheduling with the most increased cost using Equations (15) and (16). With the updated scheduling, shuffle the population and generate the new memeplex. Calculate the charging cost of each scheduling find out the minimum charging cost (X_b) and the maximum charging cost X_w^{new}.

Step 6: If the maximum charging cost X_w^{new} is less than $X_w^{current}$, calculate the charging cost of each schedule. If not, go to Step 5.

Step 7: Sort the population P in descending order according to their charging cost.

Step 8: When the number of iterations is completed, then stop the process.

Step 9: The charging schedule problem is solved by the FIFS, PSO, and SFLA algorithms, as presented in Section 4.

4. Results and Discussion

The optimal scheduling is performed for minimizing the electricity purchase cost from the PL. In Table 3, the electricity price at the 7th and 8th timeslots is high, and the low prices are at the 4th, 5th, 3rd, and 6th slots. Therefore, the PL operator can use the time slots optimally to minimize the charging cost. The charging scheduling is presented for EVs with prior reservations and for EVs arrived without a reservation. Three cases are compared, the first case represents the 20 EVs that come with an appointment (base case), the second case represents 5 EVs (such as EVs 16, 17, 18, 19, and 20) arriving without an appointment, and the third case represents 10 EVs (11, 12, 13, 14, 15, 16, 17, 18, 19, and 20) arriving without an appointment. The PL is also provided with a microgrid, i.e., the power generated from the DGs is used in the MG case whenever the MG price is less than the grid price. Furthermore, the dynamic scheduling by FIFS, PSO, and SFLA is investigated, and the results are presented and discussed with and without the microgrid scenario considered.

4.1. The Schedule Using FIFS

In general, most of the PLs use the FIFS. Apart from its simplicity (easy to be applied without optimization or decision-making framework), the main advantage of the FIFS method is that it avoids the charger being in idle mode. Table 7 presents the scheduling using the FIFS.

Each time slot's charging demand is 50.43, 61.5, 61.5, 61.5, 61.5, and 39.47 kW. The PL can assess an EV load of 61.5 kW per hour. The cost of charging all the EVs is 2442.07 €ct. The charging slot cost is 399.3, 461.0, 444.5, 424.5, 424.5, and 287.9 €ct, respectively. Even though the EVs 4, 8,11,17,19 have parking time until the 8th slot, the charging is completed before the 7th hour.

4.2. Dynamic Schedule Using PSO

The PSO technique is used to perform optimal scheduling for minimizing the total electricity purchase cost. At the beginning of each timeslot, the charging schedule for the particular timeslot is executed to achieve the minimum electricity cost. The algorithm also determines the plan for the next timeslots. However, the schedule is revised for the upcoming timeslot depending upon the arrival of EVs in the next timeslot. The average time taken to complete the charging in each time slot is 33.40, 59.99, 59.95, 59.99, 59.98, 54.30 min. The convergence curve of the PSO algorithm is shown in Figure 3. PSO's dynamic scheduling is given in Table 8, and the optimal allocation of power and resources is given in Table 9.

The EV demand is allocated to each time slot to achieve the minimum cost. The PSO's minimum cost is 2432.0 €ct, which is cheaper than the FIFS scheduling cost. As the grid's electricity cost in timeslots 7 and 8 is relatively high, PSO schedules the EVs in the first six timeslots to achieve the minimum charging cost.

Table 7. Dynamic scheduling using the first-in-first serve (FIFS) method.

ID	Demand (kW)	Timeslot							
		1	2	3	4	5	6	7	8
1	16.19	16.19	0	0	0	-	-	-	-
2	17.25	-	17.25	0	0	0		-	-
3	14.85	-	-	14.85	0	0	0	-	-
4	20.64	-	-	-	20.64	0	0	0	0
5	21.87	21.87	0	0	0	0	0	0	-
6	12.32	-	-	12.32	0	-	-	-	-
7	17.28	-	17.28	0	0	0	-	-	-
8	26.40	-	26.40	0	0	0	0	0	0
9	12.11	-	-	-	-	12.11	0	0	-
10	20.8	-	-	-	7.79	13.00	0	0	-
11	17.04	-	-	-	-	-	17.04	0	0
12	16.74	-	-	-	-	6.09	10.64	-	-
13	9.60	-	0.57	9.03	-	-	-	-	-
14	11.79	-	-	-	-	0	11.79	-	-
15	16.10	-	-	-	0	16.10	0	-	-
16	12.05	-	-	4.18	7.86	0	-	-	-
17	25.20	-	-	0	25.20	0	0	0	0
18	14.18	-	-	-	0	14.18	0	-	-
19	21.12	-	0	21.12	0	0	0	0	0
20	12.37	12.37	0	0	0	0	0	0	-
	Total	50.43	61.50	61.50	61.49	61.48	39.47	0.00	0.00

Table 8. Dynamic scheduling using PSO.

ID	Demand (kW)	Timeslot							
		1	2	3	4	5	6	7	8
1	16.19	0	16.19	0	0	-	-	-	-
2	17.25	-	17.25	0	0	0	-	-	-
3	14.85	-	-	0	14.85	0	0	-	-
4	20.64	-	-	-	0	20.64	0	0	0
5	21.87	21.87	0	0	0	0	0	0	-
6	12.32	-	-	0	12.32	-	-	-	-
7	17.28	-	17.28	0	0	0	-	-	-
8	26.40	-	0	26.4	0	0	0	0	0
9	12.11	-	-	-	-	12.11	0	0	0
10	20.80	-	-	-	12.24	8.55	0	0	-
11	17.04	-	-	-	-	-	17.04	0	0
12	16.74	-	-	-	-	6.00	10.73	0	-
13	9.60	-	9.6	0	-	-	-	-	-
14	11.79	-	-	-	-	0	11.79	-	-
15	16.10	-	-	-	0	0	16.1	-	-
16	12.04	-	-	12.0	0	0	-	-	-
17	25.20	-	-	3.11	22.08	0	0	0	0
18	14.18	-	-	-	0	14.18	0	-	-
19	21.12	-	1.17	19.94	-	-	-	-	-
20	12.37	12.37	0	0	0	0	0	-	-
	Total	34.24	61.49	61.45	61.49	61.48	55.66	0.00	0.00

Table 9. The optimal power and resources allocation using PSO.

ID	Demand (kW)	Timeslot 1			Timeslot 2			Timeslot 3			Timeslot 4			Timeslot 5			Timeslot 6		
		A	B	C	A	B	C	A	B	C	A	B	C	A	B	C	A	B	C
1	16.19	16.19	32.3	FC	0												–		
2	17.25	17.25	27.7 + 10.2	FC, MC	0												–		
3	14.85	–			0			0			14.85	44.5	MC	0			0		
4	20.64	–			0			–			0			20.64	41.2	FC	0		
5	21.87	21.87	43.74	FC	0			0			0			0			0		
6	12.32	–									12.32	15.4 + 37.4	MC, SC				–		
7	17.28	17.28			49.8 + 3.5		MC, SC	0						0			–		
8	26.40	0						26.4	52.8	FC	0			0					
9	12.11	–						–						12.11	18.8 + 8.1	FC, MC	0		
10	20.80							–			12.24	15.9 + 22.5	FC, SC	8.55	44.6	SC	0		
11	17.04	–						–			–			–			17.04	34.08	FC
12	16.74	–						–			–			6	9.4 + 15.1	MC, SC	10.73	55.9	SC
13	9.60	9.6	50	SC	0			–			0			0			–		
14	11.79	–			–			0			0			0			11.79	23.5	FC
15	16.1	–			–			0			0			0			16.1	48.3	MC
16	12.04	–						12	7.2 + 43.8	FC, SC	0			0			–		
17	25.20	–						3.11	16.2	SC	22.08	44.16	FC	0					
18	14.18							0			0			14.18	42.5	MC	0		
19	21.12	1.17	6.1	SC				19.94	59.8	MC	–			–			–		
20	12.37	12.37	37.1	MC	0			0			0			0			–		

A denotes the allocated demand (kW), B represents the time (min), and C denotes the charger where FC denotes the fast charger, MC denotes the medium charger, and SC denotes the slow charger.

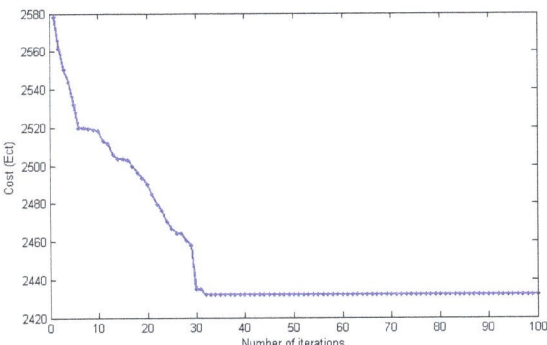

Figure 3. Convergence curve of particle swarm optimization (PSO).

4.3. Dynamic Schedule Using Shuffled Frog Leaping Algorithm (SFLA)

The SFLA is used for the optimal scheduling of EVs to reduce the electricity purchase cost. For each slot the electricity cost is 226.1, 461.0, 444.5, 424.5, 424.5, 447.5 €ct. The total electricity purchase cost is 2428.47 €ct. The optimization techniques effectively utilize the low electricity price time slots for scheduling of EVs. Also, this shows that if the number of EVs arrives with prior booking, better scheduling is obtained to minimize the grid's charging cost. The convergence curve of the SFLA is shown in Figure 4. The convergence speed of SFLA is faster than the PSO. The scheduling results obtained by SFLA is given in Table 10. The optimal power and resource allocation are given in Table 11.

Table 10. Dynamic scheduling using SFLA.

ID	Demand (kW)	Timeslot							
		1	2	3	4	5	6	7	8
1	16.19	16.19	0	0	0				
2	17.25	-	17.25	0	0	0	-	-	-
3	14.85	-	-	9.48	5.37	0	-	-	-
4	20.64	-	-	-	0	20.64	0	0	0
5	21.87	0	21.87	0	0	0	0	0	-
6	12.32	-	-	0	12.32	-	-	-	-
7	17.28	-	17.28	0	0	0	-	-	-
8	26.4	-	5.1	21.3	0	0	0	0	0
9	12.11	-	-	-	-	10.52	1.58	0	-
10	20.8	-	-	-	20.8	0	0	0	-
11	17.04	-	-	-	-	-	17.04	0	0
12	16.74	-	-	-	-	0	16.74	0	-
13	9.6	-	0	9.6	-	-	-	-	-
14	11.79	-	-	-	-	0	11.79	-	-
15	16.1	-	-	-	16.1	0	0	-	-
16	12.04	-	-	0	6.91	5.13	-	-	-
17	25.20	-	-	0	0	25.20	0	0	0
18	14.18	-	-	-	0	0	14.18	-	-
19	21.12	-	0	21.12	0	0	0	0	0
20	12.37	12.37	0	0	0	0	0	-	-
	Total	28.56	61.50	61.50	61.50	61.49	61.33	0.00	0.00

Table 11. The optimal power and resource allocation using SFLA.

ID	Demand (kW)	Timeslot 1			2			3			4			5			6		
		A	B	C	A	B	C	A	B	C	A	B	C	A	B	C	A	B	C
1	16.19	16.19	32.3	FC	0	-	-	0			-			0					
2	17.25	0	-	-	17.25	51.7	MC	0			5.37	24 + 2.3	SC, MC	0					
3	14.85	-			-	-	-	9.48	22.7 + 9.9	MC, SC	-			-					
4	20.64	-			-	-	-	-			-			20.64	9.6 + 47.5	FC, MC			
5	21.87	-			21.87	43.7	FC	0			0			0					
6	12.32	-			-	-	-	0			12.32	36.9	MC	-					
7	17.28	-			17.28	16.3 + 8.3 + 33.2	FC, MC, SC	0			0			0					
8	26.40	-			5.1	26.6	SC	21.3	42.6	FC	0			0					
9	12.11	-			-			-			-			10.52	12.5 + 33.1	MC, SC	1.58	8.2	SC
10	20.80	-			-			-			20.8			0			0		
11	17.04	-			-			-			-	41.6	FC	-			17.04	34	FC, FC, MC
12	16.74	-			-			-			-			0			16.74	25.92 + 11.34	
13	9.60	-			0			9.6	50	SC	-			-					
14	11.79	-			-			-			-			0			11.79	51.8 + 5.6	SC, MC
15	16.1	-			-			-			16.1	18.4 + 20.7	FC, MC	0			0		
16	12.04	-			-			0			6.91	36	SC	5.13	26.7	SC	-		
17	25.20	-			-			0			0			25.2	50.4	FC	-		
18	14.18	-			-			-			0			0			14.18	42.5	MC
19	21.12	-			0			21.1	17.4 + 37.2	FC, MC	0			0			0		
20	12.37	12.37	37.1	MC	0			0			0			0			0		

A denotes the allocated demand (kW), B represents the time (min), and C denotes the charger where FC denotes the fast charger, MC denotes the medium charger, and SC denotes the slow charger.

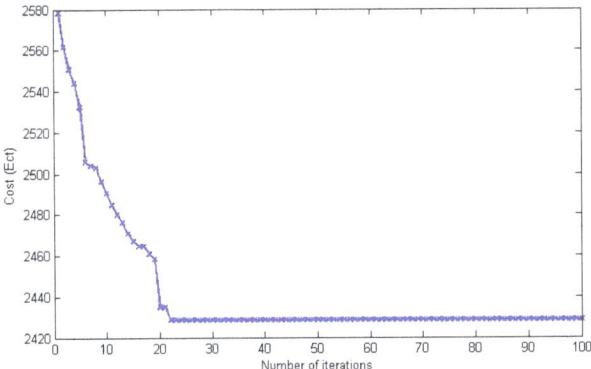

Figure 4. Convergence curve of shuffled frog leaping algorithm (SFLA).

Figure 5 shows that the optimization techniques provide a reduced charging cost compared to the FIFS charging algorithm.

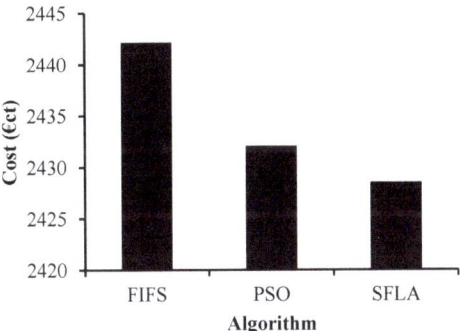

Figure 5. Cost comparison of FIFS, PSO, and SFLA.

However, the charging cost of the PL is calculated without considering the microgrid (MG) scenario available with the PL. The microgrid power can be used to minimize the electricity purchase cost of the PL. The renewable energy sources provide significant potential that can benefit the CSs. When the microgrid power is supplied to the CS, the cost is reduced significantly. Compared to the grid cost, the MG cost in all the slots is cheaper, and hence it is utilized effectively [65,66].

The optimal use of renewable energy is not only beneficial for cost reduction but also supports the grid. By using the microgrid available in the PL, the cost savings are given in Table 12. Furthermore, the charging cost of EVs with uncertain arrival is examined in the next subsection and compared with the charging cost of EVs arriving with a prior booking.

Table 12. Cost comparison of FIFS, PSO, and SFLA.

Number	Method	No Microgrid (MG) Considered	Microgrid (MG) Considered	The Difference in Cost (%)
		Cost (€ct)	Cost (€ct)	
1	FIFS	2442.0	2384.6	3.40
2	PSO	2432.0	2374.7	2.31
3	SFLA	2428.4	2371.2	2.40

4.4. Dynamic Schedule of EVs with/without Appointments

Three cases are considered in this investigation:

- Case 1: All the 20 EVs arrived with prior booking (base case).
- Case 2: 15 EVs, out of 20 EVs, arrived with prior booking.
- Case 3: 10 EVs, out of 20 EVs, arrived with prior booking.

The arrival and departure times are randomly assigned within the eight timeslots. The users with prior booking need to provide the expected arrival time, departure time, and charging demand. For EVs without booking, the EVs' expected departure time and charging demand are provided when they reach the CS. To keep the same total demand in the three cases, only the arrival and departure times are randomly generated. In each case, the results are obtained using the SFLA and then compared with the PSO and FIFS scheduling algorithms. The dynamic system can determine a charging schedule at the beginning of each timeslot using all the EVs with and without booking, and then the PL can charge arrived vehicles with the use of the schedule for the immediate timeslot. At the beginning of each timeslot, the computational process has to be executed.

It can be seen that the electricity purchasing cost of the PL is reduced when the EVs arrived with reservation. Using FIFS, the charging price for cases 2 and 3 is increased by 3.26% and 3.11%, respectively, compared to the base case. Using SFLA, the charging price for cases 2 and 3 increased by 2.06% and 3.03%, respectively, compared to the base case (case 1). Also, using PSO, the charging price for cases 2 and 3 is increased by 2.70% and 2.89%, respectively, compared with case 1. The cost comparison with and without prior booking is given in Table 13, in which the MG case is not considered. Unexpected arrival and departure are considered for unappointed vehicles, as shown in Tables 14 and 15. Table 16 shows the scheduling when 5/20 vehicles have arrived without a booking, and Table 17 shows the scheduling of 10/20 vehicles have arrived without booking. These two cases are performed to minimize the electricity purchase cost, and the results show that the charging demands can be fulfilled in the three scenarios. Also, the needs are the same for the three cases; hence the cost can be compared.

Table 13. Cost comparison of the charging schedule of EVs arriving with/without prior booking.

Cases	1	2	3
EVs Arriving with/without Booking	20/0	15/5	10/10
Cost obtained with FIFS (€ct)	2442.0	2520.7	2530.2
Cost obtained with PSO (€ct)	2432.0	2495.9	2509.6
Cost obtained with SFLA (€ct)	2428.4	2493.5	2504.4

Table 14. Unexpected arrival and departure considered for unappointed vehicles: when 5/20 EVs arrived without an appointment.

EV/SLOT	1	2	3	4	5	6	7	8
1	In			Out				
2		In			Out			
3			In			Out		
4				In			Out	Out
5	In						Out	
6			In	Out				
7		In			Out			
8		In						Out
9					In		Out	
10				In			Out	
11						In		Out
12					In		Out	
13		In	Out					
14					In	Out		
15				In		Out		
16						In		Out
17	In	Out						
18					In			Out
19							In	Out
20					In		Out	

Table 15. Unexpected arrival and departure considered for unappointed vehicles: when 10/20 EVs arrived without an appointment.

EV/SLOT	1	2	3	4	5	6	7	8
1	In			Out				
2		In			Out			
3			In			Out		
4				In				Out
5	In						Out	
6			In	Out				
7		In			Out			
8		In						Out
9					In		Out	
10				In			Out	
11					In		Out	
12						In		Out
13			In	Out				
14				In	Out			
15						In		Out
16						In		Out
17	In	Out						
18					In			Out
19								Out
20					In		Out	

Table 16. Unexpected arrival and departure considered for unappointed vehicles: schedule when 5/20 EVs have arrived without an appointment.

ID	Demand (kW)	Timeslot							
		1	2	3	4	5	6	7	8
1	16.19	16.19	0	0	0	-	-	-	-
2	17.25	-	17.25	0	0	0	-	-	-
3	14.85	-	-	14.85	0	0	0	-	-
4	20.64	-	-	-	20.64	0	0	0	0
5	21.87	21.87	0	0	0	0	0	0	-
6	12.32	-	-	12.32	0	-	-	-	-
7	17.28	-	17.28	0	0	0	-	-	-
8	26.40	-	25.208	1.192	0	0	0	0	0
9	12.11	-	-	-	-	12.11	0	0	-
10	20.8	-	-	-	20.8	0	0	0	-
11	17.04	-	-	-	-	-	17.04	0	0
12	16.74	-	-	-	-	16.74	0	0	-
13	9.6	-	0	9.6	-	-	-	-	-
14	11.79	-	-	-	-	11.792	0	-	-
15	16.1	-	-	-	16.1	0	0	-	-
16	12.04	-	-	-	-	-	12.045	0	0
17	25.20	23.438	1.762	-	-	-	-	-	-
18	14.18	-	-	-	-	14.186	0	0	0
19	21.12	-	-	-	-	-	-	21.12	0
20	12.37	-	-	-	-	6.672	5.703	0	-
Total		61.50	61.50	37.96	57.54	61.5	34.79	21.12	0.00

Table 17. Unexpected arrival and departure considered for unappointed vehicles: schedule when 10/20 EVs have arrived without an appointment.

ID	Demand (kW)	Timeslot							
		1	2	3	4	5	6	7	8
1	16.19	16.19	0	0	0	-	-	-	-
2	17.25	-	17.25	0	0	0	-	-	-
3	14.85	-	-	14.85	0	0	0	-	-
4	20.64	-	-	-	20.64	0	0	0	0
5	21.87	21.87	0	0	0	0	0	0	-
6	12.32	-	-	12.32	0	-	-	-	-
7	17.28	-	17.28	0	0	0	-	-	-
8	26.4	-	25.208	1.192	0	0	0	0	0
9	12.11	-	-	-	-	12.11	0	0	-
10	20.8	-	-	-	20.8	0	0	0	-
11	17.04	-	-	-	-	17.04	0	0	-
12	16.74	-	-	-	-	-	16.74	0	0
13	9.6	-	-	9.6	0	-	-	-	-
14	11.79	-	-	-	11.792	0	-	-	-
15	16.1	-	-	-	-	-	16.1	0	0
16	12.04	-	-	-	-	-	12.045	0	0
17	25.20	23.438	1.762	-	-	-	-	-	-
18	14.18	-	-	-	-	14.186	0	0	0
19	21.12	-	-	-	-	-	-	21.12	0
20	12.37	-	-	-	-	12.375	0	0	-
Total		61.50	61.50	37.96	53.23	55.71	44.89	21.12	0.00

In Table 13, the comparison shows that the electricity cost is less than that obtained using the FIFS scheme.

Furthermore, when the MG supplies the PL, the charging costs are reduced, as presented in Table 18. A comparison of the three algorithms' values obtained in the three cases without and with MG consideration are shown in Figures 6 and 7, respectively.

Table 18. Cost comparison of the charging schedule of EVs arriving with/without prior booking considering the MG scenario.

Cases	1	2	3
EVs Arrived with/without Booking	20/0	15/5	10/10
Cost obtained with FIFS (€ct)	2384.61	2426.01	2430.43
Cost obtained with PSO (€ct)	2374.74	2402.00	2410.10
Cost obtained with SFLA (€ct)	2371.21	2398.90	2405.30

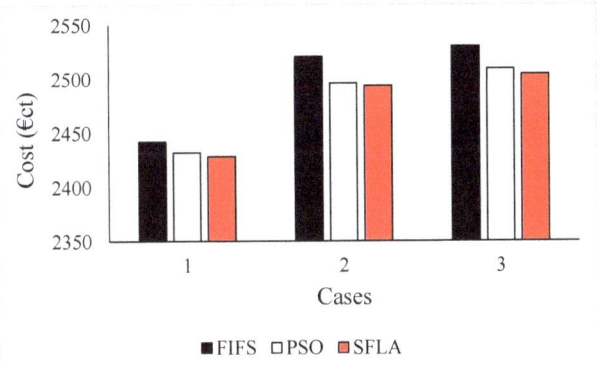

Figure 6. Charging costs without MG consideration: (1) 20/20 EVs arriving with booking; (2) 15/20 EVs arriving with booking; and (3) 10/20 EVs arriving with booking.

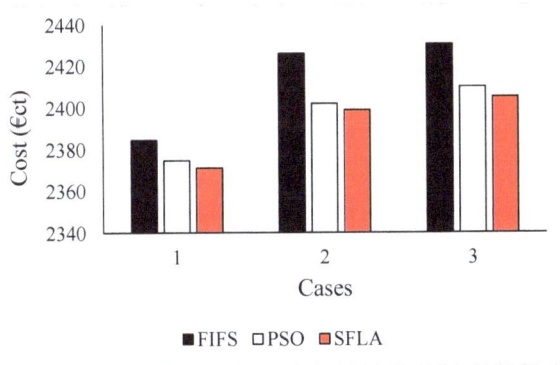

Figure 7. Charging costs with MG consideration: (1) 20/20 EVs arriving with booking; (2) 15/20 EVs arriving with booking; and (3) 10/20 EVs arriving with booking.

5. Conclusions

A dynamic charging scheduling scheme for charging EVs in a PL is proposed for minimizing the charging costs. First, a conventional FIFS scheduling scheme is performed for the EV charging. Economic scheduling is found using PSO and SFLA. In this regard, the dynamic charging scheme allocates the optimal electric power in each slot for the vehicles that have arrived. Scheduling is undertaken for the vehicles that have arrived at the PL with and without prior booking. This scheme considers the electricity price at each hour and the charging limit of the PL every hour for all the three methods considered. When the electricity price is low, the entire timeslot is fully utilized to charge the EVs effectively. Also, the PL operator provides 100% of the EV user's power within the available time. The results showed that more significant savings could be reached if the EV arrives with a prior reservation.

6. Future Works

In this work, the charging demand of each EV is considered as 100%, but this limit can vary, and random charging demand can be considered. Service delay can be regarded to benefit both the user and CS operator. Reserve charging points can be taken into account, and the charging can be undertaken for users with a special tariff based on the CS operator's decision. This kind of option available at the CS may help floating customers who want quick charging.

Author Contributions: Conceptualization, G.S.F. and V.K.; methodology, G.S.F.; software, S.K.; validation, J.S.A., Z.M.A. and S.H.E.A.A.; formal analysis, G.S.F.; investigation, V.K.; resources, S.K.; data curation, G.S.F.; writing—original draft preparation, S.H.E.A.A.; writing—review and editing, J.S.A. and Z.M.A.; visualization, V.K.; supervision, A.E.-S.; project administration, A.E.-S. All authors have read and agreed to the published version of the manuscript.

Funding: This research received no external funding.

Conflicts of Interest: The authors declare no conflict of interest.

Abbreviations

CC	Constant current
CCCV	Constant current constant voltage
CS	Charging station
CSP	Charging service provider
CV	Constant voltage
DG	Distributed generation
DR	Demand response
DSO	Distribution system operator
EPRI	Electric power research institute
EPEX	European Power Exchange
ESS	Energy storage system
EV	Electric vehicle
FC	Fast charger
FIFS	First-in-first serve
GA	Genetic algorithm
IEC	International Electro-technical Commission
MC	Medium charger
MG	Microgrid
PHEV	Plug-in-hybrid EV
PL	Parking lot
PSO	Particle swarm optimization
PV	Photovoltaic
SAE	Society of automotive engineers
SBP	SOC based priority
SC	Slow charger
SFLA	Shuffled frog leaping algorithm
SOC	State of charge
TOU	Time of use
TSO	Transmission system operator
V2G	Vehicle-to-Grid

T	Total time to charge all EVs
$C(n)$	Charging cost of EVs
$C(t)$	Total electricity purchase cost
c_1, c_2	Acceleration coefficients
C_{lim}^t	Charging limit
D_i	Demand for the ith vehicle
D_{power}^t	Power demand at time t
$E_c(t)$	Electricity price at time t
Gbest	Global best position
N	Total number of EVs
NF	Number of FCs
n_i	Status of the ith EV
NM	Number of MCs
NS	Number of SCs
Pbest	Individual best position
P_{ifc}	Rated output power of the FC
P_{imc}	Rated output power of the MC
P_{isc}	Rated output power of the SC
r	Charging time to reach 100% of the battery
R	Random distribution
R_P	Required power to charge the EV
Si	Variation in frog's location
$SOC(n)$	Power left in the nth vehicle
T_{arr}^i	Arrival time of the ith vehicle
T_{dep}^i	Departure time of the ith vehicle
V_i	Velocity of the ith particle
Xb	Best position of frog in a memeplex
Xg	Global best position
Xi	Existing position of the ith particle
X_W	Worst position of frog in a memeplex
V_c^n	Rated power capacity of the EV
ω	Inertia weight

References

1. Savari, G.F.; Krishnasamy, V.; Sathik, J.; Ali, Z.M.; Aleem, S.H.A. Internet of Things based real-time electric vehicle load forecasting and charging station recommendation. *ISA Trans.* **2020**, *97*, 431–447. [CrossRef]
2. Abdel Aleem, S.H.E.; Zobaa, A.F.; Abdel Mageed, H.M. Assessment of energy credits for the enhancement of the Egyptian Green Pyramid Rating System. *Energy Policy* **2015**, *87*, 407–416. [CrossRef]
3. Sandy Thomas, C.E. Transportation options in a carbon-constrained world: Hybrids, plug-in hybrids, biofuels, fuel cell electric vehicles, and battery electric vehicles. *Int. J. Hydrogen Energy* **2009**, *23*, 9279–9296. [CrossRef]
4. Clement-Nyns, K.; Haesen, E.; Driesen, J. The impact of vehicle-to-grid on the distribution grid. *Electr. Power Syst. Res.* **2011**, *81*, 185–192. [CrossRef]
5. Martinenas, S.; Knezovic, K.; Marinelli, M. Management of Power Quality Issues in Low Voltage Networks Using Electric Vehicles: Experimental Validation. *IEEE Trans. Power Deliv.* **2017**, *32*, 971–979. [CrossRef]
6. Zhou, X.; Lukic, S.; Bhattacharya, S.; Huang, A. Design and control of grid-connected converter in bi-directional battery charger for plug-in hybrid electric vehicle application. In Proceedings of the 5th IEEE Vehicle Power and Propulsion Conference, VPPC'09, Dearborn, MI, USA, 7–11 September 2009; pp. 1716–1721.
7. Yilmaz, M.; Krein, P.T. Review of battery charger topologies, charging power levels, and infrastructure for plug-in electric and hybrid vehicles. *IEEE Trans. Power Electron.* **2013**, *28*, 2151–2169. [CrossRef]
8. Thimmesch, D. An SCR Inverter with an Integral Battery Charger for Electric Vehicles. *IEEE Trans. Ind. Appl.* **1985**, *IA-21*, 1023–1029. [CrossRef]

9. The International Electro Technical Commission (IEC). *Electric Vehicle Conductive Charging System—Part 1: General Requirement*; IEC 61851-1; The International Electro Technical Commission: London, UK, 2001.
10. Etezadi-Amoli, M.; Choma, K.; Stefani, J. Rapid-Charge Electric-Vehicle Stations. *IEEE Trans. Power Deliv.* 2010, 25, 1883–1887. [CrossRef]
11. Akhavan-Rezai, E.; Shaaban, M.F.; El-Saadany, E.F.; Karray, F. New EMS to incorporate smart parking lots into demand response. *IEEE Trans. Smart Grid* 2018, 9, 1376–1386. [CrossRef]
12. Neyestani, N.; Damavandi, M.Y.; Shafie-Khah, M.; Contreras, J.; Catalão, J.P.S. Allocation of Plug-In Vehicles' Parking Lots in Distribution Systems Considering Network-Constrained Objectives. *IEEE Trans. Power Syst.* 2015, 30, 2643–2656. [CrossRef]
13. Su, W.; Chow, M.Y. Computational intelligence-based energy management for a large-scale PHEV/PEV enabled municipal parking deck. *Appl. Energy* 2012, 96, 171–182. [CrossRef]
14. Katic, V.A.; Dumnic, B.P.; Corba, Z.J.; Pecelj, M. Electric and hybrid vehicles battery charger cluster locations in urban areas. In Proceedings of the 17th European Conference on Power Electronics and Applications, EPE-ECCE Europe, Geneva, Switzerland, 8–10 September 2015.
15. Masoum, A.S.; Deilami, S.; Moses, P.S.; Masoum, M.A.S.; Abu-Siada, A. Smart load management of plug-in electric vehicles in distribution and residential networks with charging stations for peak shaving and loss minimisation considering voltage regulation. *IET Gener. Transm. Distrib.* 2011, 5, 877–888. [CrossRef]
16. Shao, S.; Pipattanasomporn, M.; Rahman, S. Grid Integration of Electric Vehicles and Demand Response With Customer Choice. *IEEE Trans. Smart Grid* 2012, 3, 543–550. [CrossRef]
17. Zhao, Z.; Lee, W.C.; Shin, Y.; Song, K. Bin An optimal power scheduling method for demand response in home energy management system. *IEEE Trans. Smart Grid* 2013, 4, 1391–1400. [CrossRef]
18. Richardson, P.; Flynn, D.; Keane, A. Local versus centralized charging strategies for electric vehicles in low voltage distribution systems. *IEEE Trans. Smart Grid* 2012, 3, 1020–1028. [CrossRef]
19. Fan, Z. A Distributed Demand Response Algorithm and Its Application to PHEV Charging in Smart Grids. *IEEE Trans. Smart Grid* 2012, 3, 1280–1290. [CrossRef]
20. Tan, Z.; Yang, P.; Nehorai, A. An optimal and distributed demand response strategy with electric vehicles in the smart grid. *IEEE Trans. Smart Grid* 2014, 5, 861–869. [CrossRef]
21. Xu, Z.; Hu, Z.; Song, Y.; Zhao, W.; Zhang, Y. Coordination of PEVs charging across multiple aggregators. *Appl. Energy* 2014, 136, 582–589. [CrossRef]
22. Awad, A.S.A.; Shaaban, M.F.; El-Fouly, T.H.M.; El-Saadany, E.F.; Salama, M.M.A. Optimal Resource Allocation and Charging Prices for Benefit Maximization in Smart PEV-Parking Lots. *IEEE Trans. Sustain. Energy* 2017, 8, 906–915. [CrossRef]
23. Jin, C.; Tang, J.; Ghosh, P. Optimizing Electric Vehicle Charging With Energy Storage in the Electricity Market. *IEEE Trans. Smart Grid* 2013, 4, 311–320. [CrossRef]
24. Yao, L.; Lim, W.H.; Tsai, T.S. A Real-Time Charging Scheme for Demand Response in Electric Vehicle Parking Station. *IEEE Trans. Smart Grid* 2017, 8, 52–62. [CrossRef]
25. Zhang, L.; Li, Y. Optimal Management for Parking-Lot Electric Vehicle Charging by Two-Stage Approximate Dynamic Programming. *IEEE Trans. Smart Grid* 2017, 8, 1722–1730. [CrossRef]
26. Zhang, L.; Li, Y. A Game-Theoretic Approach to Optimal Scheduling of Parking-Lot Electric Vehicle Charging. *IEEE Trans. Veh. Technol.* 2016, 65, 4068–4078. [CrossRef]
27. Su, W.; Chow, M.Y. Performance evaluation of an EDA-based large-scale plug-in hybrid electric vehicle charging algorithm. *IEEE Trans. Smart Grid* 2012, 3, 308–315. [CrossRef]
28. Kuran, M.S.; Viana, A.C.; Iannone, L.; Kofman, D.; Mermoud, G.; Vasseur, J.P. A Smart Parking Lot Management System for Scheduling the Recharging of Electric Vehicles. *IEEE Trans. Smart Grid* 2015, 6, 2942–2953. [CrossRef]
29. He, Y.; Venkatesh, B.; Guan, L. Optimal Scheduling for Charging and Discharging of Electric Vehicles. *IEEE Trans. Smart Grid* 2012, 3, 1095–1105. [CrossRef]
30. Clairand, J.-M.; Rodriguez-Garcia, J.; Alvarez-Bel, C. Smart Charging for Electric Vehicle Aggregators Considering Users' Preferences. *IEEE Access* 2018, 6, 54624–54635. [CrossRef]
31. Rotering, N.; Ilic, M. Optimal charge control of plug-in hybrid electric vehicles in deregulated electricity markets. *IEEE Trans. Power Syst.* 2011, 26, 1021–1029. [CrossRef]
32. Han, S.; Han, S.; Sezaki, K. Development of an optimal vehicle-to-grid aggregator for frequency regulation. *IEEE Trans. Smart Grid* 2010, 1, 65–72.

33. Sundstrom, O.; Binding, C. Flexible Charging Optimization for Electric Vehicles Considering Distribution Grid Constraints. *IEEE Trans. Smart Grid* **2012**, *3*, 26–37. [CrossRef]
34. Yang, S. Price-responsive early charging control based on data mining for electric vehicle online scheduling. *Electr. Power Syst. Res.* **2019**, *167*, 113–121. [CrossRef]
35. Cao, Y.; Wang, T.; Kaiwartya, O.; Min, G.; Ahmad, N.; Abdullah, A.H. An EV charging management system concerning drivers' trip duration and mobility uncertainty. *IEEE Trans. Syst. Man Cybern. Syst.* **2018**, *48*, 596–607. [CrossRef]
36. Ansari, M.; Al-Awami, A.T.; Sortomme, E.; Abido, M.A. Coordinated bidding of ancillary services for vehicle-to-grid using fuzzy optimization. *IEEE Trans. Smart Grid* **2015**, *6*, 261–270. [CrossRef]
37. Aghajani, S.; Kalantar, M. Operational scheduling of electric vehicles parking lot integrated with renewable generation based on bilevel programming approach. *Energy* **2017**, *139*, 422–432. [CrossRef]
38. Mohammadi Landi, M.; Mohammadi, M.; Rastegar, M. Simultaneous determination of optimal capacity and charging profile of plug-in electric vehicle parking lots in distribution systems. *Energy* **2018**, *15*, 504–511. [CrossRef]
39. Sedighizadeh, M.; Mohammadpour, A.; Alavi, S.M.M. A daytime optimal stochastic energy management for EV commercial parking lots by using approximate dynamic programming and hybrid big bang big crunch algorithm. *Sustain. Cities Soc.* **2019**, *45*, 486–498. [CrossRef]
40. Jiang, W.; Zhen, Y. A Real-Time EV Charging Scheduling for Parking Lots With PV System and Energy Store System. *IEEE Access* **2019**, *7*, 86184–86193. [CrossRef]
41. del Razo, V.; Jacobsen, H.-A. Smart Charging Schedules for Highway Travel With Electric Vehicles. *IEEE Trans. Transp. Electrif.* **2016**, *2*, 160–173. [CrossRef]
42. Mohsenian-Rad, H.; Ghamkhari, M. Optimal charging of electric vehicles with uncertain departure times: A closed-form solution. *IEEE Trans. Smart Grid* **2015**, *6*, 940–942. [CrossRef]
43. Xu, Y.; Pan, F.; Tong, L. Dynamic Scheduling for Charging Electric Vehicles: A Priority Rule. *IEEE Trans. Automat. Contr.* **2016**, *61*, 4094–4099. [CrossRef]
44. Rottondi, C.; Neglia, G.; Verticale, G. Complexity Analysis of Optimal Recharge Scheduling for Electric Vehicles. *IEEE Trans. Veh. Technol.* **2016**, *65*, 4106–4117. [CrossRef]
45. Wi, Y.M.; Lee, J.U.; Joo, S.K. Electric vehicle charging method for smart homes/buildings with a photovoltaic system. *IEEE Trans. Consum. Electron.* **2013**, *59*, 323–328. [CrossRef]
46. Liu, Z.; Wu, Q.; Huang, S.; Wang, L.; Shahidehpour, M.; Xue, Y. Optimal Day-Ahead Charging Scheduling of Electric Vehicles Through an Aggregative Game Model. *IEEE Trans. Smart Grid* **2018**, *9*, 5173–5184. [CrossRef]
47. Liu, Z.; Wu, Q.; Ma, K.; Shahidehpour, M.; Xue, Y.; Huang, S. Two-Stage Optimal Scheduling of Electric Vehicle Charging Based on Transactive Control. *IEEE Trans. Smart Grid* **2019**, *10*, 2948–2958. [CrossRef]
48. Sabillon Antunez, C.; Franco, J.F.; Rider, M.J.; Romero, R. A New Methodology for the Optimal Charging Coordination of Electric Vehicles Considering Vehicle-to-Grid Technology. *IEEE Trans. Sustain. Energy* **2016**, *7*, 596–607. [CrossRef]
49. Yang, L.; Zhang, J.; Poor, H.V. Risk-aware day-ahead scheduling and real-time dispatch for electric vehicle charging. *IEEE Trans. Smart Grid* **2014**, *5*, 693–702. [CrossRef]
50. Wang, B.; Wang, Y.; Nazaripouya, H.; Qiu, C.; Chu, C.C.; Gadh, R. Predictive Scheduling Framework for Electric Vehicles with Uncertainties of User Behaviors. *IEEE Internet Things J.* **2017**, *4*, 52–63. [CrossRef]
51. Chen, J.; Huang, X.; Tian, S.; Cao, Y.; Huang, B.; Luo, X.; Yu, W. Electric vehicle charging schedule considering user's charging selection from economics. *IET Gener. Transm. Distrib.* **2019**, *13*, 3388–3396. [CrossRef]
52. Chung, H.-M.; Li, W.-T.; Yuen, C.; Wen, C.-K.; Crespi, N. Electric Vehicle Charge Scheduling Mechanism to Maximize Cost Efficiency and User Convenience. *IEEE Trans. Smart Grid* **2019**, *10*, 3020–3030. [CrossRef]
53. Wang, Y.; Thompson, J.S. Two-stage admission and scheduling mechanism for electric vehicle charging. *IEEE Trans. Smart Grid* **2019**, *10*, 2650–2660. [CrossRef]
54. Zhou, Y.; Yau, D.K.Y.; You, P.; Cheng, P. Optimal-Cost Scheduling of Electrical Vehicle Charging Under Uncertainty. *IEEE Trans. Smart Grid* **2018**, *9*, 4547–4554. [CrossRef]
55. Latifi, M.; Rastegarnia, A.; Khalili, A.; Sanei, S. Agent-Based Decentralized Optimal Charging Strategy for Plug-in Electric Vehicles. *IEEE Trans. Ind. Electron.* **2019**, *66*, 3668–3680. [CrossRef]
56. Liu, S.; Etemadi, A.H. A Dynamic Stochastic Optimization for Recharging Plug-In Electric Vehicles. *IEEE Trans. Smart Grid* **2018**, *9*, 4154–4161. [CrossRef]

57. Guo, Y.; Xiong, J.; Xu, S.; Su, W. Two-Stage Economic Operation of Microgrid-Like Electric Vehicle Parking Deck. *IEEE Trans. Smart Grid* **2016**, *7*, 1703–1712. [CrossRef]
58. Taheri, N.; Entriken, R.; Ye, Y. A dynamic algorithm for facilitated charging of plug-in electric vehicles. *IEEE Trans. Smart Grid* **2013**, *4*, 1772–1779. [CrossRef]
59. Wang, S.; Bi, S.; Zhang, Y.-J.A.; Huang, J. Electrical Vehicle Charging Station Profit Maximization: Admission, Pricing, and Online Scheduling. *IEEE Trans. Sustain. Energy* **2018**, *9*, 1722–1731. [CrossRef]
60. You, P.; Yang, Z.; Chow, M.Y.; Sun, Y. Optimal Cooperative Charging Strategy for a Smart Charging Station of Electric Vehicles. *IEEE Trans. Power Syst.* **2016**, *31*, 2946–2956. [CrossRef]
61. Gan, L.; Topcu, U.; Low, S.H. Optimal decentralized protocol for electric vehicle charging. *IEEE Trans. Power Syst.* **2013**, *28*, 940–951. [CrossRef]
62. European Power Exchange. 2017. Available online: http://www.epexspot.com (accessed on 1 December 2020).
63. Papathanassiou, S.; Nikos, H.; Kai, S. A benchmark low voltage microgrid network. In Proceedings of the CIGRE Symposium: Power Systems with Dispersed Generation, Athens, Greece, 13–16 April 2005; pp. 17–20.
64. Zobaa, A.F.; Aleem, S.H.E.A.; Abdelaziz, A.Y. *Classical and Recent Aspects of Power System Optimization*; Academic Press: Cambridge, MA, USA; Elsevier: Amsterdam, The Netherlands, 2018.
65. Mostafa, M.H.; Aleem, S.H.A.; Ali, S.G.; Ali, Z.M.; Abdelaziz, A.Y. Techno-economic assessment of energy storage systems using annualized life cycle cost of storage (LCCOS) and levelized cost of energy (LCOE) metrics. *J. Energy Storage* **2020**, *29*, 101345. [CrossRef]
66. Mostafa, M.H.; Aleem, S.H.E.A.; Ali, S.G.; Abdelaziz, A.Y.; Ribeiro, P.F.; Ali, Z.M. Robust energy management and economic analysis of microgrids considering different battery characteristics. *IEEE Access* **2020**, *8*, 54751–54775. [CrossRef]

Publisher's Note: MDPI stays neutral with regard to jurisdictional claims in published maps and institutional affiliations.

© 2020 by the authors. Licensee MDPI, Basel, Switzerland. This article is an open access article distributed under the terms and conditions of the Creative Commons Attribution (CC BY) license (http://creativecommons.org/licenses/by/4.0/).

MDPI
St. Alban-Anlage 66
4052 Basel
Switzerland
Tel. +41 61 683 77 34
Fax +41 61 302 89 18
www.mdpi.com

Energies Editorial Office
E-mail: energies@mdpi.com
www.mdpi.com/journal/energies

www.ingramcontent.com/pod-product-compliance
Lightning Source LLC
LaVergne TN
LVHW070735100526
838202LV00013B/1244